Water in the Wilderness

LIFE IN THE COAST, DESERTS, AND MOUNTAINS OF PAKISTAN

Water in the Wilderness

LIFE IN THE COAST, DESERTS, AND MOUNTAINS OF PAKISTAN

Mehjabeen Abidi-Habib

Richard Garstang & Rina Saeed Khan

OXFORD
UNIVERSITY PRESS

Oxford University Press is a department of the University of Oxford.
It furthers the University's objective of excellence in research, scholarship,
and education by publishing worldwide. Oxford is a registered trade mark of
Oxford University Press in the UK and in certain other countries

Published in Pakistan by
Ameena Saiyid, Oxford University Press
No.38, Sector 15, Korangi Industrial Area,
PO Box 8214, Karachi-74900, Pakistan

© Oxford University Press 2016

The moral rights of the authors have been asserted

First Edition published in 2016

ISBN 978-0-19-940011-9

Typeset in Book Antiqua
Printed on 90gsm Matt Finish Paper

Printed by VVP

Acknowledgements
Book Design and Layout: Tehmina Rashid
Cover Design: Sabiha Imani
Photographs: Ayesha Vellani, Javaid A. Khan, Ghulam Rasool by permission
WWF-Pakistan/PWP, Ghulam Rasool, The British Library, Hasan Zaki by permission
of WWF-Pakistan and PWP, Azmatullah, Houbara Foundation International Pakistan,
WWF-Pakistan, All photographs of Katerina by African Odyssey Project,
Himalayan Wildlife Foundation, Jon Miceler, John Farrington,
Inayat Ali of Shimshal, Rina Saeed Khan, Naushaba Anjum Lahore Museum
Text: Essay and photographs in chapter *Jiwani*, 'From Iran: One People Divided by
a Border' by Green Front of Iran and Mohsen Soleymani; essay in chapter *Lungh*,
'From Siberia: The Story of Katerina' by Miroslav Bobek; essay and photographs
in chapter *Cholistan* 'From India: The Rajasthan Route' by Rahul Chandrawarkar;
essay in chapter *Deosai*, 'From Tibet: River from the Lion's Mouth, the Source of
the Indus' by John D. Farrington and Dawa Tsering; essay in Chapter *Shandur*,
'From Afghanistan: The Wakhan Wilderness' by Khushal Habibi

TABLE OF CONTENTS

MAP OF PAKISTAN

Map of Pakistan, showing the nine study areas that we call 'Water in the Wilderness'. They lie beyond the Indus River system yet are associated with life-giving fresh water, unique human communities and wildlife assemblages. Essays from neighbouring countries give a sense of continuities, both social and ecological, that transcend political boundaries being essential to an understanding of the whole region.

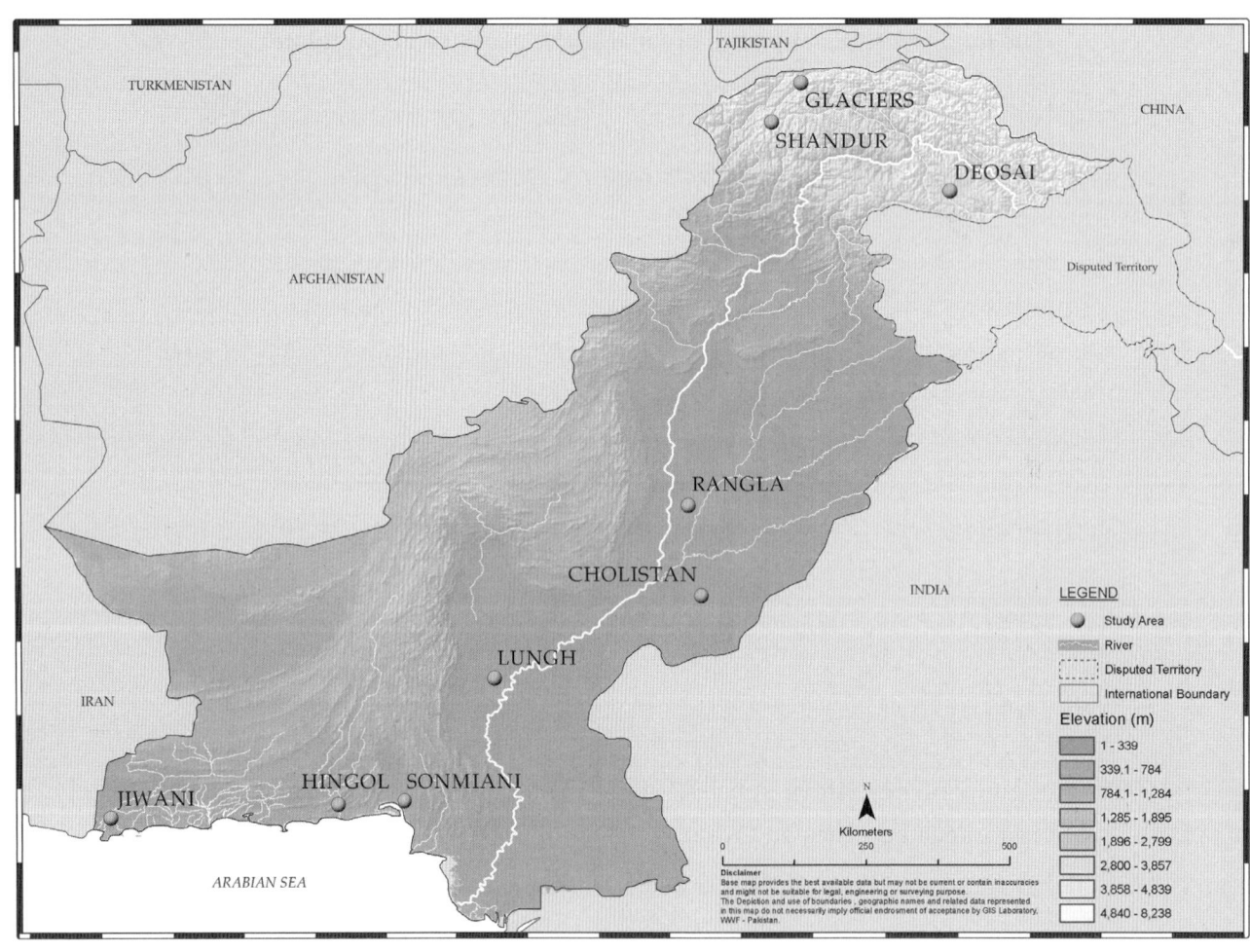

Foreword

Water in the Wilderness is a treasure. As in all great literature, we are taken to places and people we will never know and come to understand how others live. We only fully grasp the richness of a culture once we understand the deserts, pastures, mountains, rivers, and coastlines, which in turn form lives, songs, beliefs, food, and materials.

The lands of Pakistan are as diverse as the people who populate it. Mehjabeen Abidi-Habib and her colleagues show us worlds within worlds permeated by intricate webs of life that have arisen over millennia. The line between humans, plants, and creatures blends into a fusion of nature and culture, or a way of being from within, an interlinked co-evolution. A cumulative wisdom arises from the refinements of residing in one region over generations. What could be called the science of place gets coded into music, poems, rituals, and language. There are sixty-eight minor languages in Pakistan, but they are no more minor than a rare species of plant or the endangered Saker Falcon nesting in the sinuous Jiwani cliffs. Each language contains ways of knowing and relating to place that bind people and land together as a kind of living organism.

This nuanced sense of place is disappearing. Diverse habitats 'hold a key to resilience of the larger whole; and should we fail to value and protect this ecology we will collectively impoverish ourselves,' writes Mehjabeen. If we don't know where we live, the names of migratory species, or the botany of our existence, our ignorance becomes an erasure; a blotting out of who we are as people, family, and communities. When that occurs, the cultivated lands, waterways, and surrounding wilderness begin to fade away as well, exploited for their resources for people in distant lands. The natural world is the bloodstream of human culture. When living systems recede and largely disappear, we are left with modernity, soda pop signs, imported cheap clothing, depleting sugary foods, and the blandishments of the media and commerce. Although the diverse biomes of Pakistan have not yet been lost, the process of dismantling them is well underway.

We understand here the absurdity of the artificial boundaries ('unnatural barriers') drawn by colonial powers or regional conflicts. Other than reinforcing nationalism, they have no meaning for people whose cultures, rivers, mountains, and trade routes criss-cross them daily. Cultures are like birds and water. They flow and undulate across the land. Human beings move and interact with their environment in fluid ways. When those patterns are interrupted and degraded, it is like cutting off a limb.

The theme of the book is illustrated by a gorge deep inside the Hingol River watershed. There, a complex of Hindu temples contain shrines: one to the mother goddess Devi Ma and the other to Kali Ma, the goddess of destruction. The freshwater stream that cuts through the gorge is precious in this arid landscape. The temples and shrines depend on this water in the wilderness. New roads and affluence have allowed a large increase in pilgrims who visit the shrine. Due to its popularity, with both Hindu and Muslim pilgrims, the water is becoming polluted, brackish, littered with plastic, candy wrappers, and tins. More to the point, the Hingol River watershed, though located in a national park, has been proposed as a site for a large hydroelectric dam. Although the issue is now somnolent, it remains a possibility. The ancient and the new, devotion and greed, irony and loss, all converge in this story of the primacy of water and ultimately our fate.

Mehjabeen Abidi-Habib's writing is simple and exquisite. It would be easy, if not facile, for her to draw deep lines in the sand describing how venal players are destroying ecosystems and cultures. But that would not exactly be the truth. The complexity of the living systems that are beautifully portrayed here is matched by the intricacy of human beliefs, intentions, and yes, failings. Here is the heart of both: the problem and solution to the multitude of ecological crises we face in the world. They cannot be solved by righteousness or ignorance. They require a deep appreciation for all that has gone before. Virtually every living system co-evolved with pastoralists, hunter-gatherers, agriculturalists, or urbanists. Human beings created durable cultures by learning from and adapting to the flux of seasons, flora, and fauna. The most salient resource we have to address our impending losses is embedded in *Water in the Wilderness*. It is written with a compassionate and informed intelligence that honours all life on earth, including us imperfect human beings. We are gifted here by this presence of mind eloquently expressed and imaged.

What is the measure of a great book? We are not the same person when we have finished reading it. Our mind is not merely added to; our mind is different because someone has shared a way of seeing the world that was hitherto unknown to us. This is that book.

Paul Hawken
April 2015
Sausalito, California

Introduction

Imagine a mother's feelings for the child in her womb: instinctive, enfolding, boundless protection—in this book you will recognize people and places that still feel this way about being on earth.

All our ancient traditions tell us that this is the archetypal state of being here, on this earth, within nature and with each other. The ancient Hebrew word *r-ḥ-m* wends itself across millennia into Islamic concepts with the word *Raḥma* that means in Arabic 'from the womb'—a description of the quality that boundlessly enfolds all creation together into one interdependent whole. The same universal concept comes into Sanskrit as *mai-tri*, or motherly protection.

Now, we are in an age which scientists call the 'anthropocene', an era when humans have profoundly disrupted nature's rhythms with alterations, perturbations, loss, and dead zones. In response, we look within ourselves to recover our lost sense of irrevocable belonging to nature; with our innermost being. When we understand the decisive nature of the anthropocene, we seek *Raḥma* and *mai-tri* more intensely than ever before.

Where civilization has revolved around water for the past five thousand years, and a material culture associated with textiles, we, the writers, drew fresh water as one would draw a thread out of an old, intricate, enduring tapestry. A complex weave of streams, lakes, rivers, ice, and water seemed to us like silvery threads spread across this land. And we found that the people and nature of this land are bound together with water that, like a cosmic force, begets, sustains, and also destroys life.

This book is about people who still remember what it means to be necessary to the land; people without whom the land would wither. With their bodies and animals, with their clothes and language, food and memory, they belong deeply to where they are. The land itself is defined by their presence and way of life.

But to think that the people you will meet in this book are of the past would be a mistake. In fact, we repeatedly ask this question through a lens emanating from science and society. In his book

Blessed Unrest, bestselling US author Paul Hawken perceptively observes, 'we are at a threshold in human existence, a fundamental change in understanding about our relationship to nature and each other. We are moving from a world created by privilege to a world created by community.' What does community mean and how can we understand and reach out to one that is not our own? Many of us can now only comprehend this word in a loose way; our roots distantly remembered within migrations, altered villages, lost places, and cosmopolitan cultures.

The British Museum in its recent book *A History of the World in 100 Objects*, queries 'Can we ever really understand others? Perhaps, but only through feats of poetic imagination, combined with knowledge vigorously acquired and ordered'.

Water in the Wilderness uses all these three elements: the ordering of knowledge both new and old, vigour, and conspicuously, it uses our poetic imagination.

A meter of three is our book's organizing principle: three ecosystem sections to include coasts, deserts, and mountains; each section with three chapters about three different locations. Within a chapter is a string of essays about place, history, and people. Every chapter has a summary of all it offers in its introduction; read it and you get the gist. Equally, it is possible to browse any brief essay and gain a perspective.

The vigour of knowledge was much enabled by recent increments in ecosystem science. In the early 2000s, Pakistan experienced a surge of field biology that shed light on ecologies that support life in this densely populated and fertile land.

You will discover, as the cultures of the ancient Indus Valley civilization knew, that fresh water is at the heart of society. The recent scientific surveys explain the interconnections between wetlands, or bodies of fresh water that occur anywhere outside the sea. This wetland science offers new understandings of rivers, lakes, marshes, tributaries, floodplains, deltas, and of course, those magnums of water reservoirs, glaciers.

Our colleague, Richard Garstang, a naturalized Pakistani of South African origin, was a pioneer of this new ecology, positioned as he was in the country's premier conservation organization, WWF-Pakistan. Simultaneously, an era of impactful new institutions emerged—in the making for at least twenty years prior to the science. Countrywide, innovative collectives of people in villages and cities, national and regional initiatives had arisen to address issues shared as a commons, the realm of my research. Rina Saeed Khan, who reports as a journalist on socio-environmental issues, is an expert at communicating new happenings in this field, winning awards for her brave investigations of remote places and vulnerable people.

Across fifteen years, the three of us have worked together on many things. With *Water in the Wilderness*, we mounted our most daring venture yet: to incorporate new science, the immediacy of fieldwork and historical perspectives for insight into social and ecological change unfolding in Pakistan. We journeyed all over Pakistan for some four years, variously hosted by army encampments, government officials, non-governmental organizations, road-building projects, and people in small hamlets. The book has taken a decade to prepare.

Sweeping through landscapes we saw big changes driven by future economies like the Gwadar Port, or by climate shifts that impact glaciers; tiny changes like the overuse of Peelu tree roots in the Cholistan desert; and marvelled at the mysterious shells lodged in desert forts that betray ancient, forgotten rivers. These directed us to questions about what to expect from changes that we are collectively unleashing.

Information gathered in *Water in the Wilderness* is ordered across space and time, both, new and old. Our historical research in the British Library and Pakistan's public libraries brought fascinating perspectives of colonial annals carefully documenting and mapping the conquered lands of India.

Clearly, Pakistan's national boundaries are a recent phenomenon. Only if we let ourselves fly across its boundaries, as do the migratory birds recurrent in our story, do we truly begin to understand nature and culture's continuties. So we commissioned five fascinating trans-boundary essays, or written pieces by people deeply involved with history, landscape, and culture from beyond the borders of Pakistan.

Exposing remarkable similarities with Jiwani in the Makran coast, an essay from Iran shows the people of Gwatar Bay on the other side of the border. In Sindh, Lungh Lake is a sanctuary for Siberian birds that fly to the Arabian Sea in winter. Our riveting story from the Czech Republic tells how a Siberian black stork named Katerina was radio-collared in her faraway summer home and tracked through her flight and plight down the Indus Valley. We learn of the seamless connections between families of the divided Cholistan and adjoining Bikaner in Rajasthan. From Afghanistan's Wakhan, a distinguished writer describes his first-hand account of the nomadic Kuchi and Kirghiz who roam there and their routes of passage once connected with Shandur, a high altitude wetland of Pakistan. Finally, we reach into Tibet, that great plateau and source of all Asia's life-giving rivers, to read of the source of the Indus itself from the eyes of a Tibetan anthropologist and an American wildlife specialist.

Water in the Wilderness is conspicuously full of poetic imagination; there would be no other way to deeply enter into the world of others.

Shaded in grey, our cameo pieces tell the slightly fictionalized, sometimes hybrid stories of real people whom we met. For me, these characters are iconic of each place and spirit. To Richard we owe the indomitable chieftain Talal of Hingol, who braves the sands with his camel, face scarred by a scimitar, whose name means dewdrop. High up in the Himalayas, to Richard also we owe the story of Abdul Wahid of Astore, hunting wild ibex with chivalry of times immemorial: retaining fear and reverence for the life he is about to take. Rina tells the story of Kulsoom, the Ismaili schoolteacher of Gwadar's old town, remembering it as a bustling port of the Arabian Sea, and another sketch of the last Prince of Mastuj, erstwhile gallant polo player and owner of the high altitude peat lands of Shandur as his royal legacy. My renditions include the old storyteller of Jiwani and the little boy, Khuda Dad of Bhera, who dreams of a computer in a village with no electricity.

I am particularly proud of the unique photographs taken mostly by Ayesha Vellani, whose eye for people within landscapes, their lives and emotions, are infused with haunting beauty. Ayesha, a young mother herself, can get closer to the mother and child motif than any photographer I know in Pakistan; a society where only a woman is allowed so near this sanctum of the inner family.

Javed Ahmed Khan and Ghulam Rasool are also among the country's most admired nature photographers and they completed our dream team who capture the beauty of all we convey like no one else I know. Their love of nature and all things natural spurs them to physical ventures few other women and men would dare. Our graphic designer Tehmina Rashid's meticulous layout and respectful handling of visuals resonate with the character of the work itself.

Over four years, Richard, our technology-savvy field biologist, planned all our expeditions with me. While he stayed behind at his day job, our team of almost all women went into remote hinterlands. What befell such an unlikely team in a country like Pakistan is in itself a series of adventures. Three women, a baby on the hip, and a howitzer cannon-sized telephoto lens raised many eyebrows as we travelled.

Soon after an international airing of a surreptitious video of Osama bin Laden, a friend advised that we switch the location of our field study site. Geologists in the US had surmised the rock cliffs filmed behind Osama were just those featured in the wetland site of our destination. 'You are sure to be spotted by satellite surveillance and baffle the analysts at your peril,' the friend suggested. Ayesha's baby girl Zoya and husband Jawad travelled on several expeditions with us; the baby's feeding times monitored through walkie-talkie as her mother photographed migratory birds at Lungh Lake in Sindh. In the Cholistan desert, we arrived late one night to meet our host, a retired army Colonel. He was flummoxed when I said we ought be out in the field at the crack of dawn—4.30 a.m., when the desert light is low on the horizon. It took a stern phone call from Richard, once himself in the South African army, to convince our host, with words I still do not know, that we are a team who say what we mean.

Carrying a voice from Pakistan's wilderness, *Water in the Wilderness* speaks to us. Like the prophet Isaiah's voice crying in the wilderness, it foretells a profound possibility. It whispers how we might understand each other and our role in a world created collectively.

Growing up in Europe, I gathered that it took 2,000 years to draw up cultural stereotypes hinting that North European culture is brutish and staid, while Southern Europe, where I lived, is creative and joyful. Just such stereotypes are threatening our identities once again, dividing and separating, when we have been given that great opportunity: to enter the world of others and become infused with boundless wonder and the urge to protect—embrace archetypal *Rahma* and *mai-tri* that flow effortlessly when we begin to understand the other, and in so doing, realize our own higher possibilities.

Mehjabeen Abidi-Habib
Lahore, April 2015

SECTION ONE

The Coast

Introduction

Scholars say that culturally, Pakistan is a country that faces both East and West at one and the same time. One of the western-most faces of Pakistan is to be found in the little town of Jiwani, the last habitation of any significant size along the Makran coast. Here, a new 'Great Game' is soon to be played out, as China prepares to use the nearby port of Gwadar as a place of transit for its goods; and oil and gas may be transported from Central Asia through this warm water port. For now, Jiwani is a peaceful seaside anchorage located on one flank of the crescent-shaped Gwatar Bay. Straddling the political boundaries of two countries, Gwatar Bay is half in Iran and half in Pakistan.

A place of renowned sunset vistas over luminous expanses of the Indian Ocean and remembered by travellers in their annals of centuries past, Jiwani seems an abandoned place of natural beauty, of historical mementoes, and a lively intermingling of cultures where man-made political boundaries become blurred. Set among these natural and cultural continuities, Jiwani may soon be transformed by two port towns: one in Pakistan and one in Iran. The port city of Gwadar in Pakistan will offer a deep sea port and investment haven to the global commerce community; while the port and warehousing facilities of the Iranian Chahar Bagh are ready with advanced infrastructure and trading facilities. In Gwadar, however, unrest is in the air as the local Baloch resent the displacement of their people and the acquisition of land for the new development projects. They are angry with the government for using their local resources without taking their rights into consideration.

At the heart of Gwatar Bay is the Dasht River, whose basin area is 116,065 square kilometres, spread over the arid hinterlands of the central Makran hill range and the Kuh-i-Bampusht hills in the Iranian Sistan region. Sweet-water from the Dasht River flows into the Pakistani side of Gwatar Bay as if through an artery that sustains all life in this forbidding and dry terrain. Along its meandering course from north to south, the river creates little lakes and rock pools where people, animals, and plants have eked

(facing page)
Jiwani anchorage, where a moonrise illuminates the last habitation of the Makran coast, beyond which lies Iran.

3

Sweet-water from the Dasht River creates little lakes where people eke out a sustenance in the arid hinterlands.

out a sustenance. The mouth of this river at the Arabian Sea fans out into an estuary that combines its own sweet-waters with the warm, salty ones of the Indian Ocean: an ideal place for mangrove forests, and all the terrestrial and aquatic life that teems in this rich ecosystem.

Along the mud-flats and mangroves of the Dasht estuary, more than a hundred and twenty-five species of migratory birds have been sighted. In springtime, the eye scanning the shoreline is met with the profile of flocks of endangered Dalmatian Pelicans who overwinter here from Siberia. The falcon, a winter bird from the steppes of Mongolia and Kazakhstan, also calls upon the rocky cliffs around Jiwani for the winter months. Along these cliffs are pristine golden beaches where the only footprints might be those of feral dogs. The dogs scour the beaches for baby hatchlings of the Green Turtles when they struggle out of their sandy shoreline nests and clamber towards the vast blue ocean beyond. Upon occasion, the skeletal parts of whales are seen along these same shores, washed up like flotsam; a telltale sign of the route of passage that these large mammals take as they feed and travel along the Arabian Sea coast.

Off the coastline of Jiwani, what lies beneath the ocean waters is still a mystery to scientists. The land-forms that we now see were once areas submerged in the ocean, and rose up to expose beds of coral life, now dead and dry. One can walk over blankets of hard, spiky grey coral near the roads of Jiwani wondering if this is what lies under the ocean, in a colourful and vital form. Although this area of natural wonders is recognized by the International Ramsar Convention as a site worth protecting for nature conservation—a 'Ramsar site'—the national protection status given to Jiwani remains unclassified, so that the area is vulnerable to exploitation and unwise developments that can harm and destroy irreplaceable natural resources.

Like the whale bones, there are vestiges of built heritage strewn across Jiwani and Gwadar. These are the reminders of the region when it was successively used by the Portuguese, the Omanis, and the British as a vantage-point for controlling the passage of ships

*Delicate baby Green Turtle hatchlings struggle towards their vast
Indian Ocean home as feral dogs scour the beach to hunt them.*

from the Arabian Gulf to India. The Americans also used Jiwani airport as they refuelled aeroplanes here during the Second World War, and remnants of the old barracks still dot the landscape. Along Jiwani's shoreline, tall blockhouses stand like abandoned multi-storied buildings, now empty shells of look-out towers dotting the seascape; memories of a time when armies protected this maritime gateway. In Gwadar, the historical Ismaili quarter is testimony to the time when this was a thriving coastal town where slaves and spices from Africa were traded by Omani sultans. A sixteenth-century Portuguese watchtower, complete with a rusty cannon, still stands guard over the old harbour adjacent to the Ismaili quarter.

The people of Jiwani are deeply connected to Iran, more so perhaps than to the country of which they are citizens. They are a mixed ethnic race of Irano–Baloch settlers who are thought to have descended from the north-western stretches of Iran a thousand years ago. For centuries, men from this part of Makran have been recruited into the Omani 'Baloch Regiment' as fearless warriors and mercenaries. This connection has permeated the folklore

Victoria Hut is a charming and carefully preserved colonial structure still used by the Pakistan Coast Guard.

The sleepy border bazaar of Jiwani may be transformed by the large port and warehousing facilities of nearby Gwadar and Chahar Bagh in Iran.

The people of Jiwani are a mixed ethnic race of Irano-Baloch
settlers who came here a thousand years ago.

and language of the local people. Trade with Iran is a humming business that brings in daily provisions to the grocery shops of the main street of Jiwani.

Yet, change is in the air in Jiwani as its fate is tied to that of Gwadar. In this chapter, we ask those questions that venture beyond the heady promises of development projects and the new 'Great Game'. Will Jiwani, with its Dasht estuary and deeply traditional people who have witnessed change slowly and organically, be overwhelmed with the fast-paced future of Gwadar Port? Will these cultural and ecological continuities be understood and protected as vital to this part of the Makran—or are we to witness the jettisoning of a fragment of what the British recognized 'as one of the remotest races and places in the world'?

A relic block house used by the British as a vantage point for controlling the passage of ships from the Arabian Gulf to India.

The pristine Daran Beach in Jiwani is Pakistan's largest marine turtle nesting sites.

Reaching Ecology's Edge

Beyond Jiwani is a kind of Irano–Pakistani empty-quarter, a land of the great, salty dryness of the Sistan-Balochistan deserts; so that the little anchorage of Jiwani near the mouth of the Dasht River is a place on the far edge of a range for many animals and plants. It is here that the falcon roosts on rocky seaside cliffs after its long migrations from Central Asia, that the Dalmatian Pelican arrives in winter; and the passages of turtles and whales take place within the marine underwaters.

Jiwani is a place where the edges of Pakistan blur into a grey fusion of wilderness and humanity that is somehow both valuable and vulnerable. This delicately fringed selvedge holds a key to the resilience of the larger whole; should we fail to value and protect this ecology, we will collectively impoverish ourselves.

The key to understanding the ecology of Jiwani's delicate thread of freshwater and the natural diversity it nurtures is to imagine the whole interdependency of nature in this location as one connected piece of differing, but interlocking segments.

The coast of Jiwani consists of stone shelves, mud-flats, and dense mangrove forests which are ideal feeding and roosting grounds for migratory birds. The arid and semi-arid areas found inland are home to desert life, of which conservationists place particular importance on the Marsh or Mugger Crocodile. In addition, recent studies show more than 125 bird species recorded from different habitats in the region.

In Gwatar Bay, there are about 20 square kilometres of mangrove forest which exists along the bay in the trans-boundary region. The mangrove forest of *Avicennia marina*, the predominant species prevalent in Jiwani, extends into Iran. The pouch-shaped Gwatar Bay lies between the headlands of Iran and the rocky platform of Jiwani, bordered by a swampy region, which is the delta of the Dasht River. The river depends upon rain, hence it floods the area only once in two or three years. The total area covered by the bay is about 250 square kilometres—half of which lies in the Iranian territory and the remaining half in Pakistan.

*Rocky seaside cliffs around Jiwani mark places where migratory falcons from
Central Asia roost after their long journey.*

Forming a botanical zone named the South Iranian Rocky Uplands, this hardy tree species forms part of the Chish-lad forest adjacent to the mangroves.

Owing to the infamously extreme temperatures of this region, its vegetation consists mainly of grass, hardy spiny scrubs and trees forming a botanical zone known as the South Iranian Rocky Upland Formations. Acacia, Prosopis, and Tamarix trees are common, and with a small patch of Acacia forest known as 'Chish-lad' located adjacent to the mangrove forest near Dirr village. Another small patch of Tamarix is located south-west of the mangroves, close to the Dasht River estuary, whilst endemic species of annual herbs, such as a type of violet known as *Viola Makranica*, form an ephemeral ground cover that comes and goes with moisture. Precious grass and palm species are used extensively by the local people for housing, ropes, and livestock feed.

On the cliffs surrounding Jiwani are the resting and feeding places of two falcon species; raptors that migrate from Siberia during the winter and reach here hungry and tired, staying until they return to their breeding grounds in spring. The Saker and Peregrine falcons are both used in the three-thousand-year-old tradition of falconry that originated with the Central Asian nomads and exists still as the symbol of masculine aristocracy in the Arabian Peninsula. The falcon remains the national bird of the United Arab Emirates, and it is to this destination that thousands of the birds are smuggled to be sold at large prices. Although the Peregrine is one of the world's most widespread raptors, the Saker Falcon is an endangered species that—with its large wingspan—hunts in open grasslands and cliffs by horizontal pursuit of rodents and birds. Some seven to eight thousand of these birds were thought to survive in the wild in 2004. Although accorded legal protection in Pakistan, falcon trapping takes place annually on surrounding cliffs where umbrella-shaped mud hides with viewing holes are used to lure the birds with doves, pigeons, and quail as bait. Every year, many birds are confiscated and set free from Pakistan's territories, but many are also smuggled, enabled by lucrative bribes and profits.

East of Jiwani is its other valuable and vulnerable natural endowment: four golden beaches that are Pakistan's largest marine turtle nesting sites. The Green Turtle (*Chelonia mydas*) and the

Olive Ridley (*Lepidochelys olivacea*) are both marine turtles of the warm tropics, and it is on Jiwani's beaches that some of their major global nesting sites are found. Although both are carnivorous as adults, eating shrimp, sea urchin, and jellyfish, they also eat algae and seagrass when young, and this dietary array is found around the waters of Jiwani. To survive, both these endangered species need the protection of their nesting sites, which elsewhere in the world are destroyed by beachside developments. Here in Jiwani, it is the feral dogs that most threaten the otherwise pristine beach nesting grounds of both species. The Green Turtle is the best studied species by scientists, and we know that an individual can live up to eighty years in the wild, but that only one per cent of the hatchlings survive up to maturity. The major nesting sites of the

This tiny baby Green Turtle leaves its nest to swim into the ocean for the first time, and may live as long as eighty years of age.

Indo–Pacific subpopulation are located in Oman, Karachi, Jiwani, and nearby Astola Island.

The Marsh Crocodile is another important reptilian species found in the Jiwani area, where it has a small population. A medium-sized crocodile, it has the broadest snout of any living member of this genus, which is mainly confined to the Indian subcontinent. It occurs in water ponds in remote areas of the Dasht River, about 90 kilometres north of Jiwani. The Dasht River teems with wildlife such as crabs, snakes, and lizards, and along its mouth one can see flocks of migratory pelicans in the distance. Amongst these are the highly endangered Dalmatian Pelicans, distinguishable by their prominent beaks and yellow chrysanthemum-shaped head plumage.

Among the marine mammals, Humpback or Plumbeous Dolphins are commonly observed offshore and the Black Finless Porpoise has also been reported from the area. Scientific observers report more than twelve species of cetaceans along this coast, including both whales and dolphins. Fin Whale strandings have been recorded twice, in 2003 and 2009, and usually happen when they are trapped in gill-nets during fishing expeditions. Pakistan's Jiwani Conservation and Information Centre of the WWF (World Wildlife Fund-Pakistan) has received a whale skeleton which is displayed for educational purposes.

A skeleton of the Fin Whale, washed up on shores as it swims past the Makran coast, is displayed in the Jiwani Conservation and Information Centre.

Hair Coral washes up on the shores of Jiwani, hinting at the rich underwater coral life on reef slopes 10 to 30 metres deep.

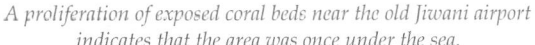

A proliferation of exposed coral beds near the old Jiwani airport indicates that the area was once under the sea.

*From Siberia, highly endangered Dalmatian Pelicans, with their legendary beaks
and chrysanthemum-shaped head plumage, flock to freshwater near Jiwani.*

The proliferation of dead and exposed coral beds and different shells found in Jiwani indicates that the area was once undersea. In particular, the warty coral and table coral that are found in abundance just below Jiwani's soil usually exist on the upper edge of outer reef slopes 10 to 30 metres deep. In fact, the world's largest bivalve was dug up near Jiwani's old airport. The area around the old airport, which was used actively during World War II by the Allies, is littered with old coral oyster beds. Recently, a navy contractor was intercepted by conservationists as preparations were being made to use the fossilized coral found in the area as building material for the road.

For its ecology, Jiwani is undoubtedly a place of many wonders that have been recognized within conservation circles. This has resulted in several legislative and international measures being put into place to secure its future. These include the declaration of the Jiwani Wetland Complex as a Ramsar site, which is an international treatise that binds countries to protect and monitor wetlands of global significance. The Daran Beach has been declared a wildlife sanctuary as a turtle nesting site of significance by the Balochistan Government. Fishing trawlers along the coast are under increasing positive pressure to use cetacean-excluding devices to protect whales and porpoises.

Nonwithstanding these significant protections, Gwadar's possible hyper-development in the medium-term, the ongoing Baloch insurgency, and the oil smuggling across Iran into Pakistan remain risks that could disturb the delicate inter-connectedness of this ecological edge, which is both pristine and fragile.

A fisherman throws his net wide into the fertile waters of the Makran coast.

Rocky seaside cliffs around Gwadar are the hallmark of this untamed coastline.

The mangrove forests near Kollani provide wood for house building,
which is seen as a task that is undertaken communally.

From Iran: One People Divided by a Border*

The border between Pakistan and Iran, like all political boundaries, is an unnatural barrier. For centuries, people on both sides have had similar customs and traditions, an almost identical way of life, and the same dependence on natural resources, such as the dense mangrove forests that are found in the estuary of the Dasht River, which forms the border. The pouch-shaped Gwatar Bay lies between the headlands of Iran and the rocky platform of Jiwani. It stretches from Iranian territory into Pakistan: the Arabian Sea on one side, and on the other, people who care little for political confines. They are more concerned with the rigours they must face to eke out a life in a land prone to both floods and drought.

The mangrove belt that stretches into Pakistan belongs also to them. About 100 kilometres from the Iranian port of Chabahar, and near enough to the border with Pakistan to allow the people to cross over in comfort, is the village of Kollani. The border region is plagued by poverty, and Kollani lacks access to basic needs like clean water, healthcare, and education. These problems are complicated by the shortage of livelihood opportunities, hence people find it hard to make conserving the nearby mangrove forests a priority.

The settlement of Kollani came into being at the turn of the twentieth century when settlers from Gwatar Bay came together to make a village. Due to a lack of amenities, Kollani was almost a failed town, and many villagers were thinking of moving back to the bay area to be closer to the sea and fishing when a severe flood devastated the region. A long drought followed and some moved back to the bay to fish; to compensate for farming and livestock breeding. Others chose to remain due to their historical ties to Kollani, which a hundred years of tradition aided in cementing.

In the close-knit village of Kollani, the rich live in cement houses and the poor in clay dwellings fortified by mangrove wood. The nomads, some of whom have now become permanent settlers,

---

*Text and images adapted from the document entitled, 'Community Empowerment for Mangrove Conservation in Guatter Bay, Iran', by the Green Front of Iran.

A glimpse of Kollani Village in Iran, where an identical way of life shows customs and people similar to Jiwani.

live in huts. They are considered outsiders, although they have been in the area longer than the village has been in existence.

The people of the area are Muslims and speak Balochi just like their neighbours in Pakistan, with much the same style of dress. Most of the men are able to speak Persian, but only educated women can, and there is a dearth of them as people are reluctant to send their daughters to school.

The people of Kollani, and of Gwatar Bay itself, have strong ties with communities on the other side of the border. Many have relatives there, and crossing the border for a social visit is common. On the Iranian side, the custom is that visitors to a prominent village elder's home can help themselves to food from the kitchen. This is a known method of hospitality.

Due to floods and droughts, agriculture is no longer viable in Kollani. Fishing can be done for seven months in the year, but the lack of access roads makes it impossible for fishermen to sell directly to the market. They are thus dependent on middlemen, with the result that some fishermen prefer to take their catch across the border to Pakistan to sell, where they seem to get more favourable prices. There is also better regulation of fishing practices on the Iranian side of the border during the spawning season and hence more fish. As a result of this enforced control of fishing, Iranian waters yield richer catches and Pakistani fishermen often venture into their waters to fish.

One significant source of income for the local community is the trafficking of kerosene and gasoline across the border to Pakistan. However, this is a major threat to the mangroves and marine life in the border area due to oil spills. The hurried transfer of smuggled oil in barrels from one speedboat to another results in oil spills into the sea which degrade its water quality and adversely affect marine life. The pollution also results in the mortality of mangrove saplings.

Kollani, as the community closest to the area's rich mangrove forests, has long benefited from them. The forests have provided wood for fuel, house building, and cattle fodder. The villagers say that if it had not been for the mangrove leaves, they would have lost their camels during the long drought. The local people claim that the seeds of the mangrove forests in their area were originally brought over from Pakistan more than three hundred years ago.

The mangrove plant called 'timmer' locally — known by the same name given to the mangroves in Pakistan — is treated with respect. Concern about its gradual decline is evident; rural communities often know exactly what the problems are, but need help in arranging solutions. A Non-Governmental Organization (NGO) by the name of Green Front, formed in 1989 in Tehran, provided invaluable assistance to the villagers of Kollani. With technical assistance from WWF-Pakistan, the Green Front implemented a community-based project to conserve the local mangrove forest near Kollani from 2002 to 2004. The project successfully motivated the local community to manage their natural resources, and raised awareness about the threats facing the mangroves located on various islets, mud-flats, and tidal lagoons behind Gwatar Bay.

On the Pakistan side of Gwatar Bay, the mangrove forest covers around 21 square kilometres and is no longer exploited for fodder or fuelwood by the local communities. With help from WWF-Pakistan, they too have realized that the mangroves are a sanctuary for valuable marine resources. The Gwatar Bay area is considered to be of outstanding ecological importance as it not only supports mangroves, but is also home to endangered marine turtles, marine mammals, marsh crocodiles, and a variety of migratory birds.

*Kollani's rural men, assisted by NGO staff, are seen here prioritizing solutions
to local problems by a system of ranking them with pebbles in a grid.*

In the silvery Makran moonlight, Haji Ibrahim, the storyteller of Jiwani, weaves his ancestor's folklore into tales of the smooth continuities of the human heart and the flight of birds.

The Storyteller of Jiwani

Only once in a blue moon, the storyteller of Jiwani sits down in a friend's courtyard to recite his entrancing stories. Interwoven with folklore of his ancestors, Haji Ibrahim's stories are poetic narrations in Balochi–Persian, of the loves, the travels, and the natural beauty of his native land. The stories straddle what is now Iran and Pakistan, following no boundaries, but telling of the smooth continuities of the human heart and the flight of birds over centuries past.

On cotton-filled floor mats and Irani ottoman cushions, he sits within a cluster of his little audience. The silvery Makran moonlight glimmers on his age-softened facial contours and his pearly white turban; his eyes fixed on the distant horizon, hands in motion, he relates with a soft, lilting flourish in rhyming couplet,

> *An Alexandrene parakeet once settled upon a beautiful*
> *cage in an opulent courtyard and asked the caged bird*
> *within: Why do you mourn so? The lonesome caged bird*
> *replied: 'I mourn for I am a migratory bird, and my*
> *winter home do I pine for; the dry twigs of my land are*
> *to me more precious than the lush greens of my prison*
> *home...*

Haji Ibrahim was once a recruit to the Sultan of Oman's Baloch Regiment, a fearsome and hardy group of warrior-mercenaries who left Makran to serve in the principality. Now, retired and reticent, he sails across the Arabian Gulf from Jiwani every year in his own vessel and collects his pension from the coffers of the Sultan. He remembers fighting the 1956 war in Aden when tensions arose between the Aden Protectorates and the United Kingdom which had long controlled the Gulf of Aden as a passage to India.

His couplet captures the longings of the Baloch recruit abroad:

> *A Baloch once laid down his rifle at the feet of his*
> *Muscati paymaster saying that he would return home.*
> *When asked why, he replied that he longed for his home,*
> *the call of the seesee partridge, the sussurating babul*
> *and jand trees...*

When Haji Ibrahim was a little boy, he used to sit cross-legged in the gatherings of the elders of his village, nodding off as night wore on. But in his young mind, the lilting poems of the old storytellers stayed on. He grew fascinated by them, and slowly began to memorize the centuries-old repertory of culture and folk tradition as his hobby. Now, the echoes of those voices fade away forever, as he observes that his grandchildren no longer have the patience to listen to folk stories, and no longer connect to the traditional rites. They don't know Arabic and Persian as he does, and forget their daily prayers frequently, seduced by the fastness of disco music and English-medium city schools.

As he watches the times changing and his old age ripens, he sees a future that he cannot be responsible for and hardly sees any part in; a time which will defy the traditional wisdom that his ancestors had cultivated, and that he now embodies. He knows of the land speculators who comb the Makran coast to procure cheap property and sell to hopeful holidaymakers and resort developers. 'One day,' he says, 'my people will be like the Palestinians, dispossessed of even the graveyards of their own ancestors...' A note of inevitable sadness creeps into the greying eyes and his voice fades away as the moon sets in the silvery horizon.

In Jiwani, a little boy from what was once, 'one of the remotest races and places in the world'.

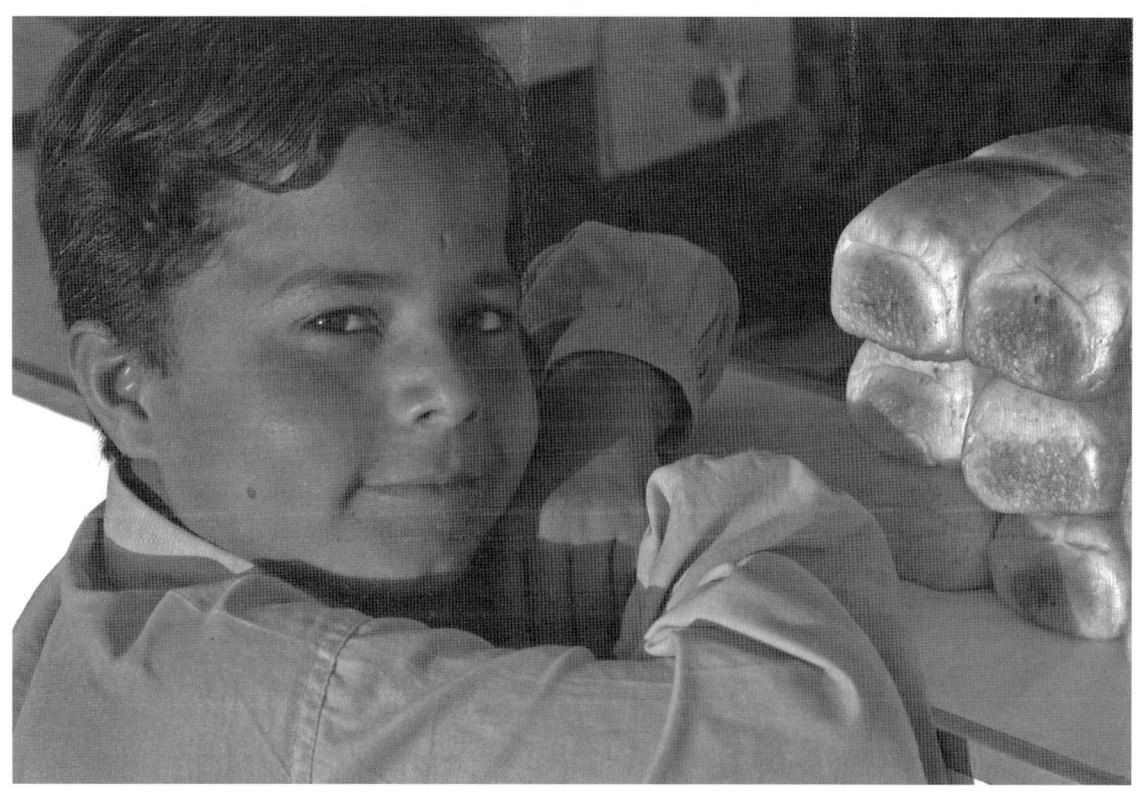

A girl in old Gwadar whose face reveals the Zanzibari connections of her ancestors.

Gwadar's Elusive Gold Rush

Jiwani has long been protected by its very isolation. The pristine beaches, rich marine life, and coral beds as yet remain intact. Will the economic tide that is coming to Gwadar adversely affect the precious ecology of Jiwani? Completed in 2008, Gwadar Port has transformed a sleepy Baloch fishing town into a major deepwater port. China is now set to take over the operation of this strategically located port, situated just 624 nautical kilometres to the east of the Strait of Hormuz, a major oil shipping lane.

The name Gwadar is derived from two Balochi words: 'guad' or wind, and 'dar' meaning gate, and literally means 'gateway of winds'. The hope is that Gwadar will eventually become a gateway to the oil and gas riches of Central Asia. From Gwadar, Jiwani is just 70 kilometres away on the Gwadar–Jiwani–Gabad road that was completed in 2009. The road provides motorists on the Makran Coastal Highway direct access into Iran.

When the deep-sea port was under construction from 2000 to 2008, Gwadar, which is located on a peninsula jutting into the Arabian Sea shaped like a hammerhead, was full of signs proclaiming upcoming new holiday resorts and housing colonies. Land speculation abounded and local residents described it as a gold rush, with property prices skyrocketing. Investors even attempted to unsuccessfully bid for Jiwani's scenic Daran Beach, home to the endangered Green Turtles who lay their eggs on the shore.

However, once the seaport and the Makran Coastal Highway were completed, trouble spread to Gwadar in the form of the fifth Baloch insurgency. Baloch nationalists view Gwadar as an imposition by the federal government and are opposed to development in the area because they say these mega projects will marginalize the local Baloch population. Government troops and installations across the province, including Gwadar town, came under rocket attacks and bombings. In reprisal, dozens of Baloch activists were picked up and tortured; their bodies dumped on the Makran Coastal Highway. The Balochistan Liberation Army, a new militant group thought to be associated with the insurgency, was banned by the Pakistan Government and branded a 'terrorist organization'.

The Gwadar Master Plan was approved by Balochistan's provincial government during the term of President General Pervez Musharraf (1999–2008). It provided space for residential, commercial, recreational, and industrial activities in Gwadar. The private sector was encouraged to invest in almost all these activities by the Gwadar Development Authority, the regulatory body in charge of the city's growth. The Gwadar peninsula is nearly 310 square miles across, with approximately 50,000 inhabitants, mostly fishermen. Many of its beaches are already polluted by diesel oil used in the engines of most of the fishing boats in Gwadar. Jiwani's rich coastal marine life may well be affected by any pollution emitting from Gwadar in the near future.

Already, illegal oil smuggling from across the border in Iran has become a noticeable environmental problem in Jiwani. This oil is brought in by road or from across the sea. The sea route is very risky because of frequent incidents of large oil spills into the jetties and creeks where the boats unload the oil. Every day, the launches bring in hundreds of barrels of oil, which are unloaded onto vehicles headed for the interior of the country while the spillover is dumped into the water. These spills are affecting the dense mangroves found in the border area; the oil also enters into the food chain and some fish found here have an oily odour.

According to the Pakistani government, the future hope for Balochistan province now lies in the development of Gwadar town and the expansion of the deep-sea port. Yet Gwadar has always had a contentious history. For almost two centuries, it was owned by the Sultan of Oman. In 1781, the Khan of Kalat granted an exiled prince of Muscat the revenues of Gwadar as maintenance while he lived on the Makran coast. The prince later returned to Oman and became the Sultan, but Gwadar did not revert to Kalat, although the Khan asked for its return. In 1839, the British conquered Kalat and became its suzerain power. After Pakistan came into being in 1947, the Pakistan Government once again took up the question of the ownership of Gwadar. Finally, Pakistan's Prime Minister Sir Firoz Khan Noon entered into negotiations with the British, which resulted in Gwadar being instated in Pakistan in 1958. Today, one can see the Sultan's old fort, now a police station, and some of the old Omani-built watchtowers.

The most historic part of Gwadar is the Ismaili quarter, the oldest Ismaili settlement in Pakistan. This was the original Gwadar, before it spread out and started developing. The well-educated Ismailis, who are followers of the progressive-minded Aga Khan, have lived in Gwadar for almost four generations. They moved here from Africa, probably via Zanzibar, the prosperous spice and slave-island which also once belonged to the Sultan of Oman. The Ismaili quarter looks similar to the inner city of Zanzibar with its multi-storied white buildings with wind-catchers and wooden balconies. The Aga Khan Trust for Culture has already helped to conserve the old stone town in Zanzibar and there is hope that they might undertake a similar project in Gwadar to protect its cultural heritage in the future.

An old Portuguese watchtower in the Ismaili quarter is said to date back to the sixteenth century, and near this tower is the Ismaili Jamaat Khana (congregational place), which was built in 1805. One can see the Portuguese influence in the architecture as the building has been well maintained over the years. The entire Ismaili quarter, although very narrow and winding, with crumbling old houses, has proper sanitation and all the houses receive piped water. The Ismaili community have pooled their resources to maintain the Ismaili quarter in Gwadar.

The old port town of Gwadar has for centuries also been a smuggler's paradise. It was once infamous for its human trafficking in slaves and today it is a place where illegal immigrants are smuggled into the Middle East and beyond. Now, the idea is to capitalize on its location at the mouth of the Persian Gulf for trade in oil. China helped build the port through the state-owned China Harbour Engineering Company, and the Chinese Government has already invested more than USD 220 million on the project. A group from Singapore was then awarded a 40-year lease to run the port, but this has been rescinded and the port will be taken over for operation by China. As an emerging superpower hungry for energy, China needs access to the oil and gas reserves of the Central Asian states. The Chinese are also keen to assist Pakistan in building other roads to acquire a 3,500 kilometres road link between Kashgar in their western Xinjiang region and Gwadar.

Some economists had predicted that the completion of the Makran Coastal Highway that now links Karachi to Gwadar and Jiwani would herald a new phase of rapid economic development for the region. This has been stalled, however, by the violent Baloch uprising which spread to Gwadar in 2009. With China now taking control of the Gwadar Port, it appears that there is a long-term vision for Gwadar that will eventually be turned into reality. China sees Gwadar as the gateway for its products to the international markets and senses that the port can play a major role in serving as a corridor for energy, cargo, and services among Central Asian countries and the Gulf states and beyond.

Gwadar's main economic activity, for now at least, remains fishing. On Gwadar's main beach, fishermen are still building their wooden boats and preparing to come back to shore after a day's fishing: in the years to come, this will change as a city rises in the desert and migrants flock in to take advantage of new economic opportunities. While the Ismaili quarter might thrive in this modern new era, what is the future for these uneducated fishermen? Will they be forced to give up their livelihoods and move into squatter settlements? Close by lies the town of Jiwani, with its bay full of rich coral beds and extraordinary marine life. Who will protect Jiwani from the onslaught of development? What happens to Gwadar will affect Jiwani in the years to come.

At the port of Gwadar, shipbuilders still work in the old way, but
transformation into a modern deep-sea port means change is in the air.

The head of Gwadar's Ismaili 'Jamaat Khana' stands outside this congregational place.
Many residents have migrated away, leaving behind shuttered, padlocked buildings.

Gwadar's Old Ismaili Quarter

During the era of mercantile activity on the Makran coast in the eighteenth and nineteenth centuries, many Ismaili families moved to the growing trading port of Gwadar. The Ismailis are followers of the Aga Khan, their living Imam, and are an offshoot of the Muslim Shia sect. They are a small but influential minority who traditionally excel at trade and commerce. Many have established their businesses in Gwadar. The Ismaili quarter is a small enclave of the tightly-knit community, located near Gwadar's old harbour. A little-known architectural gem, melding the East African Islamic building vocabulary with Portuguese influence, the narrow pedestrian streets of the Ismaili quarter are a reminder of what Gwadar once looked like. However, in the last few years, many of the Ismaili residents have left—today most of the tall, white houses are shuttered and padlocked, and the narrow alleyways under the latticed balconies are empty.

Kulsoom Qurban Ali, a middle-aged schoolteacher who lived in the Ismaili quarter for most of her life, was one of the last to leave. 'This is the oldest part of Gwadar,' she said, pointing to a tower like fortress in the square outside her house. It is a formidable looking Portuguese built tower made of beige coral rock bricks. A narrow flight of stairs leads to the first floor, a musty room covered with the litter of centuries. In the middle, still facing a narrow window looking out to the sea, lies a rusting old metal cannon, a reminder of Gwadar's strategic position in the maritime trade from the West to India. A wooden ladder leads to the top floor of the Portuguese tower, from where there was a commanding view of the harbour beyond. The Portuguese held Gwadar briefly in the sixteenth century, and established a trading post when it was an old fishing village.

There is another fortified tower inside the Ismaili quarter, Omani in character, with its circular Islamic design made of mud plaster. This tower was built more recently, probably in the nineteenth century, when the Sultans of Oman had consolidated their rule over Gwadar and wanted to protect the flourishing slave and spice trade from Africa to the Middle East and India. Kulsoom

A little-known architectural gem, the narrow pedestrian streets of the Ismaili quarter are a reminder of what Gwadar once looked like.

recalled that during her childhood Arab prisoners were sometimes brought and jailed in the Omani tower near her home.

'I grew up across the courtyard in that house over there,' she gestured towards a large three-story house with wooden balconies. Gwadar's Ismaili quarter is crowded with multi-storied houses with filigree wooden balconies and rooftop wind-catchers. Kulsoom estimated that over six hundred Ismaili families once lived in the quarter, but many left when Gwadar was returned to Pakistan in 1958. Around three hundred Hindu families living here also left in 1958, although, they were offered Pakistani citizenship by the government.

Over the years, many Ismaili families have moved to Oman or Dubai or Canada. Only a few dozen Ismaili families are left in this old settlement. Many members of Kulsoom's family have also moved abroad. 'They moved because there is no higher education available for their children here and livelihoods are dwindling.'

Kulsoom's ancestors came from Bombay, settling in Africa and finally arriving in Gwadar. 'I would say that at least three to four generations of Ismailis have lived here in Gwadar. My husband's family were here for the past 250 years.' Kulsoom's home is sunken below street level, and was full of old furniture, including ship's trunks and maritime fittings from Bombay. A skylight let in sunshine during the day. She thinks it dates back to the eighteenth century. The Jamaat Khana is nearby and it is a large and spacious building, where the remaining members of the Ismaili community come together every night for the *maghreb* (evening) prayers.

Having spent her childhood years in Gwadar, Kulsoom has been a witness to dramatic changes in the old coastal town. She vividly recalled the time when there were no vegetables grown in Gwadar and only fish was available as the staple diet. 'In those days, it would be difficult even to get to Karachi. We would have to wait for a boat to take us there.' She also remembered the shortage of fresh water when growing up, which is still a problem for the town. In her view, 'it is important for Gwadar to go forward with the development of the port and coastal highway. I think change will come slowly ... not overnight.'

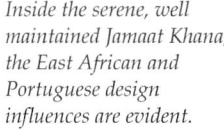

Inside the serene, well maintained Jamaat Khana, the East African and Portuguese design influences are evident.

Introduction

Only 120 kilometres north-east of the metropolis of Karachi flows a lone river, forgotten in time. The Hingol River meets the Arabian Sea in a gentle estuary that is a pristine natural area almost untouched by humankind. Along its meandering course through rugged ravines and mountains, the river collects water from streams and aquifers that run through one of the world's driest regions. Hingol is a perennial river whose waters flow throughout the year; it has a huge catchment area, estimated to be twice the size of Switzerland.

The Hingol River rises many miles to the north, and flows through one mountain range after another to the south until it enters the sea. The coastal strip is the only segment of the Hingol River that is penetrable to human traffic. For centuries, the mouth of the river has been used by camel transporters plying their way from Sindh to Iran. A few fishing villages are the only settlements along the way. The mountains that rise behind the river in great folds of dry sediment have been eroded by wind into dramatic land

(facing page)
The Hingol River meets the Arabian Sea in a gentle estuary that is a pristine natural area almost untouched by humankind.

Graves like these are the stuff of legends. It is locally believed that these date back to Muhammad bin Qasim's presence in India at the turn of the last millenium.

formations that resemble great cathedrals, one aligned in front of the other. Many compare this dramatic and hostile landscape to the Grand Canyon National Park in the United States.

This land and seascape was declared Pakistan's largest national park in 1988. What appears to be an area of vast emptiness is in fact a unique assemblage of highly adapted life forms coexisting without human interference. Underfunded and unappreciated, the national park should be one of the top priorities for conserving natural resources, whose aesthetic and economic value to the country is beyond estimation. Since it was first studied in detail by biologists in 1996, Hingol National Park has emerged as the most significant example of desert biodiversity that exists in this part of the subcontinent.

Deep inside a gorge along a tributary of the Hingol River is the revered Shri Hinglaj Mata Temple complex. Hinglaj, as this area is called, is believed to be the western-most point of the sacred world of the Hindus. For thousands of years, this gorge and its surrounding landscape features, which include mud volcanoes, rock cliffs, and natural springs have been sacred to Hinduism, one of the world's oldest living religions. Pilgrims from all over the subcontinent come here on a physically arduous journey every year in April. Hindu sacred tradition has it that the genesis of divinities are associated with specific areas, and that these are essential centres for the community to derive lessons, strength, and unity with regard to Creation. The world over, locations like Hinglaj are becoming recognized as sacred landscapes not to be defiled, but treated as an essential part of social identity. Mount Fuji in Japan and Ayers Rock in Australia have won special protection as sacred places to be preserved for posterity.

In the midst of this untouched wilderness, a national controversy has erupted over the construction of the proposed Hingol Dam. New schemes for dam construction in Pakistan have generally won votes for politicians and helped them to stay in power. But recent wisdom on the dam issue at the global level advocates options that can reduce, or at least delay, the need for dams, particularly when they are proposed inside national parks. In 2002, the World Bank

A nomadic Khosa Baloch girl's smile belies the journey through adulthood that is as rugged as the landscape itself.

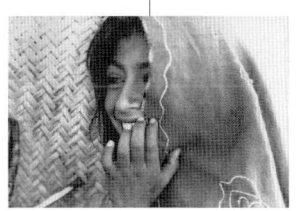

offered a USD ten million grant to the Government of Pakistan for the protection of three biodiversity hotspots, including Hingol. Sensitized by the work of the World Commission on Dams, the World Bank threatened to withdraw the grant unless the Hingol Dam plan was dropped. Faced with this choice, the Balochistan Government has shelved the plan for the dam. Had the dam reservoir been built it would have flooded parts of the sacred Hinglaj landscape and changed the ecology of the river estuary.

In 1998, Pakistan began the implementation of a major coastal highway linking Karachi to a new port town of Gwadar. This highway runs through the Hingol estuary, opening up the area to the world for the first time in history. The road will bring many uncertainties for the ecosystem as well as sudden new freedoms and changes for the inhabitants of Hingol.

In this chapter, we offer some subtleties involved in under-standing this great wilderness, and suggest that with thoughtful development, there are opportunities for making Hingol an example of a holistic protected area that can be a national park as well as a sacred site for a Hindu minority community within Pakistan and the larger community elsewhere. Such a landscape would be an environmental and political asset to the country.

Hingol includes a portion of sea, and is therefore the only terrestrial and marine national park in the region.

*Beyond the barrier of the coastal mountain range, shimmering slivers of life-giving water
are the only evidence of the Hingol River in the dry season.*

Hingol National Park

Approximately 120 kilometres west of Karachi lies the Hingol National Park. At 6,190 square kilometres it is the largest national park in Pakistan, although most visitors would not know that since there are no signs or check-posts telling them that they have now entered a national park. More importantly, however, Hingol includes a portion of sea, and is therefore the first terrestrial and marine national park in the subregion.

The most striking feature of Pakistan's Makran coast are the rows of mountains that run into the sea. These are part of a larger set of mountains that are a result of a geological event that started almost 130 million years ago. At that time, the continents of the earth consisted of two land masses, one in the south and the other in the north. The southern landmass started to break up and a large fragment which we now know as the subcontinent of India began moving northwards and collided with the Asian plate. This massive collision generated by the uplift a series of mountains which we know as the Himalayas, the highest mountains in the world, and just adjacent to them lies the Karakoram range. There are a whole series of mountain ranges that run down to the Arabian Sea, and this is how the mountains of the Hingol National Park were generated.

The wildlife of Hingol National Park owes its diversity to the great range of habitats. The eastern part of Pakistan consists of a massive plain which is now inhabited by almost 120 million people, and just a few million more occupy the rest of the mountain ranges. The Makran coast is one of the least populated parts of the Earth. The desert has protected the wildlife and there are eight distinct habitat types in Hingol.

Starting on the southern side of Hingol, the first habitat one encounters is the sea coast and this stretch runs for some 100 kilometres. Scientists studying the wildlife on the beach during an ecological survey conducted in the summer of 1999 discovered many relics that indicated the type of animals existent in the sea that forms part of the Hingol National Park.

Around eight dolphin skulls were recovered, and after analysis, it was found that there were at least three, perhaps five types of dolphins that exist in the marine waters around Pakistan's Hingol National Park. In addition, two whale skeletons were found and the whale skulls indicated that they were two distinct species.

Of particular interest to the scientists who studied the beaches were the two species of marine turtles that were found. Both Green Turtles and Olive Ridley Turtles use the beaches of Hingol as nesting grounds and although they are quite common locally, they are very rare in the rest of the world and are considered endangered species.

Hingol's shoreline is a mixture of sandy beaches and rocky portions, dramatic cliffs that run off directly into the sea, and the combination of these landscapes generate a great diversity of niches which can be occupied by wildlife. At one point, known as Boji, there are a series of caves that actually reach right down to the high water mark.

In Hingol National Park, the seaward margin of the coastline is dominated by unconsolidated deposits of sand and extrusive mud erupted from numerous volcanoes, which are conspicuous features of the Hingol coastal region. The presence of active mud volcanoes in the park clearly indicates that the area is still tectonically active. In the south-eastern part of the park, active mud vents are found near Chandragup and at the western part of Ras Kachari.

Perhaps 25 per cent of Hingol's surface area is a huge flat plain that lies between the shoreline and the mountains. Consisting in part of totally barren areas which are covered by salty mud-flats and other areas that are covered by mini dunes, what might be called a micro dune habitat. These micro dunes are particularly interesting owing to their diversity of life forms. Mini sand-dunes are fixed on the windward side by a small bush or a clump of grass which dampens wind speed and results in sand deposition just behind the bush or clump of grass, and thus the micro sand-dune forms. It provides a habitat for wildlife, a perch for birds, a place where reptiles can hide out, even a mound in which burrowing mammals can make themselves a nest. There are several small mammals associated with these micro dunes: the long-haired desert hedgehog whose tracks were found almost throughout the region, and a number of small jumping mice or gerbils, as well as a range of reptiles.

The extensive salt pan is covered by dried, brackish mud. This salt pan is so level and so hard that one could quite easily land a large aircraft on it. As expected, there is very little in the way of life on this extensive salt pan, but there is enough to support the odd predator like the Sooty Falcon, which was spotted in Hingol — only the second sighting in eighty-two years in Pakistan of this special bird of prey.

The Hingol River rises many miles to the north and flows successively through one mountain range after another to the south, until it enters the sea at almost the centre of Hingol National Park. At that point it forms a large estuary, an area where the sea water comes in at high tide and flushes out at low tide, to form a vital breeding ground for fish and other forms of wildlife that exist in the region. In fact, the estuary is one of four vital fish breeding grounds that support the fisheries of the Makran coast. Several specimens of marsh crocodiles spotted here are regarded as an

Successive parallel mountain ranges rise up as if crumbling cathedrals in an ancient land.

*The awestruck faces of the children of Kund Malir village, the first generation
to have road access to Balochistan and the metropolis of Karachi.*

A fisherman's lifestyle on the Makran coast is largely a matter of watching and waiting
for the weather and water conditions that bring in the fish.

endangered species and not found further to the west of Hingol National Park.

Also associated with the river valley is a range of vegetation that supports birdlife which would not naturally have occurred in the more arid desert parts of the park. Branching off the main river valley are a series of small river valleys or ravines and one of these is of particular interest because it has been observed as a Hindu shrine for a substantial period of time. The shrine is called Nani Mandir and researchers have found a diversity of wildlife in this ravine that did not occur anywhere else, including two species of fish and several frogs and crabs. It was on the north-facing hillsides adjacent to the valley in which the shrine occurs that one can still find Ibex. The Ibex population apparently prefer the upland portions of the park and favour the steep north-facing slopes of the mountains. As one gets to the lower mountain ranges, closer to the sea, a different species of ungulate can be found. In this region, one finds the Urial, the wild sheep that live in the desert portion of the Makran coast.

In all, a hundred and seventeen species of summer birds have been recorded at the Hingol National Park. There are many other migratory birds which enter the region during the winter months. Amongst the summer residents are White Cheek Bulbuls, snipes, Egyptian vultures, and osprey. In the large expanses of mud-flats adjacent to the estuary, a whole range of wading birds occur, particularly lapwings, stilts, little plovers.

Despite the massive size of the park, only 2,000 people permanently reside inside the area and perhaps another 2,000 are associated with its borders. This makes it one of the least populated areas on the face of the earth. The people living in the small villages like Kund Malir and Sapat Bandar inside the Park are hardy and well adapted to the harsh terrain, quite unaccustomed to dealing with visitors from the outside world. They seem focused, quite intently, on the business of survival.

When Hingol first drew the attention of biologists twenty years ago, they immediately realized they were dealing with a great range of biodiversity. First gazetted in 1988, the original area covered by the park was 1,650 square kilometres. Subsequently, Hingol National Park emerged as the most significant example of desert biodiversity that exists in this part of the subcontinent. Based on concern for the integrity of habitats, the southern boundary of the Hingol National Park was extended to include the Hingol estuary and offshore marine habitats to a depth of five fathoms. As a result, the new boundaries of the Hingol protected area now encompass an estimated area of about 6,190 square kilometres. Hingol National Park lies under the jurisdiction of the Balochistan Forest Department and a comprehensive management plan remains to be put into place.

Hingol National Park's most valuable quality is its unbroken wilderness; an area that has not been developed and has potential for carefully planned ecotourism. When we look to the future and plan what will become of this national park, its great asset must be foremost in planning efforts. The park plans should not in any way destroy its wilderness character. Wilderness, after all, will soon become one of the rarest commodities on the face of this earth. Hingol is Pakistan's most worthy candidate for development as a 'wilderness park'.

Deep inside a gorge near the bank of the Hingol River is the revered Shri Hinglaj Mata Temple complex.

Hinglaj

Deep inside a gorge near the bank of the Hingol River is the revered Shri Hinglaj Mata Temple complex. Before the construction of the coastal highway, it would take almost eight hours on a four-wheel vehicle from Karachi to reach this ancient site of pilgrimage for Hindus located in the Pubb mountains of the Kirthar range. Now, with the road completed up to Aghor, it takes four hours to get to Hinglaj—a distance of 267 kilometres from Karachi. The journey into the narrow rocky gorge where the temples are located has to be taken by foot, following a gentle stream strewn with large boulders and bordered with vegetation. The towering cliffs on both sides of the gorge close in, cutting off the sunlight and producing an eerie stillness punctuated only by the cries of birds and the occasional ringing of the temple bells. Unlike most traditional Hindu temple complexes, there are no elaborate stone carvings or intricate wooden pillars to be seen—this is a 'nature shrine', devoted to Devi Ma (mother goddess), who is the creator of all life. The shrine proper is only a deep recess inside the gorge that has been enclosed by three walls. There is another smaller shrine, merely an altar, for Kali Ma, the goddess of death and destruction. Hence, the two most powerful forces of life and death are worshipped alongside, in the heart of a ravine that has been blessed by fresh water—a

The Hindu shrine itself is only a deep recess inside the gorge, enclosed by three sides of a rock cavern.

rarity in an arid region considered one of the least-populated and developed areas on earth.

The gorge is a sanctuary not only for religious devotion—both Hindus and Muslims can be found praying peacefully together—but for wildlife. On the high ridges surrounding the ravine, visitors have spotted the Sind Ibex, while the freshwater stream inside the ravine (fed by a small pond higher up at the other end of the gorge) contains fish, frogs, and crabs. But unfortunately, the water is turning brackish, and plastic bags and tins can be seen wedged between the rocks. The building of the Makran Coastal Highway has meant that more visitors and pilgrims can now access this remote area. According to Arjun Das Lasi, who is the chief organizer of the Hindu organization called the Hinglaj Sehva Mandli which arranges an annual pilgrimage to Hinglaj, almost six thousand devotees were expected in 2002 for the five-day festival. 'We expect pilgrims from all over Pakistan and as far away as Canada and Kenya,' he said. 'For a fee of one thousand rupees per pilgrim, we arrange housing, transport, and food for the five days of the festival.' The pilgrims come every April to pay homage to the mother goddess. They seek to have their sins cleansed and their spiritual wounds healed in the natural setting of the shrine and its environs.

The festival comprises several elaborate rituals and includes Muslim participants. As one Muslim devotee explained, 'It is only Allah who can help me and answer my prayers, but I believe that the Kali Ma is very powerful and I come to Hinglaj to ask for her intervention. She helps all those whose hearts are clean—regardless whether you are Muslim or Hindu.' Some Hindu devotees liken their pilgrimage to Hinglaj to the Muslim pilgrimage to Mecca and Medina.

The first officially organized festival was held in 1972. Since then, each year at a declared date sometime in spring, the pilgrims gather at the Swami Narayan Temple in Karachi, from where they proceed together to Hinglaj aboard trucks, pickups, and jeeps. It is a festive occasion, with plenty of music and singing, as the devotees say special prayers for a safe and fulfilling pilgrimage.

'... the celebrated Hindoo temple of Hinglatz dedicated to kalee ... to which many thousand pilgrims yearly resort.' Henry Pottinger, Travels in Beloochistan and Sinde, 1816.

After the partition of the subcontinent, the Hindu minority community has kept a low profile in Pakistan and the large festival is not even mentioned in the local media. But pre-Partition Hinglaj was a famous destination for pilgrims across the subcontinent. Rajput kings would regularly come to the site from Rajasthan and Gujarat on camels; difficult journeys that would last for months. The famous Sikh Guru Nanak also visited the shrine and described it in his writings. The British have also mentioned the temple in their travelogues—Lieutenant Henry Pottinger writes in *Travels in Beloochistan and Sinde*, published in 1816, of 'the celebrated Hindoo temple of Hinglatz, dedicated to Kalee ... to which many thousand pilgrims yearly resort'.

The Muslim locals of the area often refer to the bigger, enclosed shrine as 'Nani ka Mandir' (Nani's Temple). 'Nani' is yet another name for the Devi Mata, the most powerful of the female goddesses who is the protectress of her devotees from evil.

The colourful myths surrounding the origins of Hinglaj Mata date back to times immemorial, when gods and goddesses were said to rule the universe. According to one legend, there was once a cruel king called Hingola, descended from the Mongol clans, who became furious over his brother's death at the hands of the god Ganesh and unleashed a reign of terror over the people of this region. The people prayed to Devi Mata to rid them of the cruel king, who was so powerful that it was said that his death could occur only in a place where sunlight could not penetrate. Devi Mata cornered Hingola inside a dark cave in Balochistan and killed him with her trident, the *trishul*. In his dying moments, the king prayed that in future this place would be named after him and that it would become a site for prayer and worship.

Another legend tells the story of Shiva, who carried the immolated body of his divine consort Sati all over the land. Her skull with its *hingula*, or vermillion mark on the forehead, fell in the gorge where the shrine is located—hence the name of the place came to be known as Hinglaj ever since. To this holy spot which was sanctified by the visit of Rama, Sita, and Lakshman, the great Sindhi Sufi saint-poet, Shah Abdul Latif (1689–1752), also retreated in the company of *jogis*, or ascetics. The status of the mother goddess is closely connected with the worship of Shiva, but enjoys far greater reverence as the cult centre of his consort. The various forms in which the goddess appeared at different times and places are known by various names—Kali, Durga Chandi, and Hinglaj—and represent distinct personalities to their worshippers.

Today, pilgrims (called *kapris*) to Hinglaj Mata begin their spiritual journey by wearing a braided necklace (*thumra*) and shaving their heads at Aghor in a ritual of cleansing and purification. In fact, the first step of the purification process is undertaken by stepping into the Hingol River at Aghor. Here, special chants are recited as devotees cleanse themselves ritually with the river water for their pilgrimage ahead. Their next stop is at a small shrine on the way to the gorge where they perform more ritual prayers (*puja*) to the idol of the elephant god Ganesh (the slayer of the cruel king's brother). According to another myth, Ganesh's head, which was cut off by his father, was also found in Hinglaj. It is also said that 'Hing' means stone and 'laj' means honour. Another myth records that the god Rama also visited Hinglaj to beg forgiveness from Devi Mata after killing a Brahmin. Rama performed his *puja* at the ocean shore, and pilgrims used to go to the mouth of the Hingol River to perform this ritual, but it was removed from the festival to save time (the ocean is some distance away). Pilgrims are asked to bring coconuts, incense, and cloth (*chunri*) with them as offerings to the various gods and goddesses.

Throughout the pilgrimage, devotees chant and sing *bhajjans* (devotional songs) in praise of Devi Mata and in recognition of her *shukti* (power). Aside from performing *puja* at the various shrines, there is an interesting ritual performed at the Devi Mata's temple inside the gorge. Here, below the temple, there is a tunnel accessed by two small wooden doors. The tunnel is considered to lead into the womb of the Devi Mata deep inside the rock walls, and those pilgrims who want to become brothers first cleanse themselves in the stream and then crawl through the tunnel together. They emerge as 'spiritual brothers', united by their journey into the womb. The water in the spring inside the gorge is also considered to be holy, and pilgrims often wash and bathe themselves in it before performing the rituals at the temples.

Later, the pilgrims burn coals and pray for world peace for almost four or five hours at a stretch, before visiting other natural sites outside the gorge. These include the Uneel *Kund* (well), which is said to be an unfathomable abyss dug by the goddess, and the Chaurasi *Pahar* (mountain) which they climb to save themselves from rebirth. They say that pilgrims can free themselves from 84,000 life cycles by climbing the mountain. At the top, they perform the ritual of making small stone houses, symbolic of their homes in the afterlife.

Of all the rituals performed in these natural settings, perhaps the most dramatic is the visit to the active mud volcanoes located on the way to Aghor. The larger volcano, a triangular white shape looming up from the desert that is reminiscent of the Great Pyramids, is called Chandragup and it is said to be the head of the god Shankar. Here, pilgrims climb up the steep mud volcano and toss coconuts into the bubbling crater to appease the god. It is one of the most obvious manifestations of the worship of natural phenomena during the festival.

In the wilderness of Hinglaj, nature is an extension of the divine that has been imbued with mythological significance. Pilgrims to Hinglaj undergo these difficult rituals to cleanse their sins and deepen their faith in their gods, who symbolize the power of the natural world. Unfortunately, the pollution and disturbance caused by their large numbers is threatening to spoil the very essence of what they are worshipping as sacred—the pure and natural environment that is a manifestation of the divine.

Hindu pilgrims at Hinglaj undergo difficult rituals to cleanse their sins and deepen their faith in their gods, who symbolize the power of the natural world.

The Hingol mountain range enfolds in its arms a large, flat basin that lends itself well to the building of a reservoir.

The Hingol Dam Controversy

The world over, big dams are no longer the icons of prosperity and development that they were perceived to be some twenty years ago. Decades of civil protest against the harmful effects on people and ecosystems have made their dent on global policy on large dams, hence, development agencies are now very cautious before going through with these mega projects. But in Pakistan, the romance with large infrastructure projects, including dams, still shimmers in the air like the promise of an oasis in the desert, as politicians mouth salvation for the poor and contractors continue to skim off profits from building works.

A controversy over Hingol National Park and the proposed dam within the protected area has been brewing throughout the 1990s, only to be shelved at the turn of the century when the World Commission on Dams released its global findings to the public. Yet this hornet's nest has still not entirely been smoked out of existence as the political climate continues to be turbulent in the country and leaders look around for assets to win over public opinion.

Hingol was declared a national park in 1988, a status which ensures that it is preserved as an area of outstanding ecological value to the nation both now and in the future. According to Balochistan law, this status indicates that there will be a minimum of human activity in the area. Specifically, no cultivation or mining is allowed here; as is forbidden the hunting of animals, collecting of artifacts, felling of trees, grazing of domestic livestock, and polluting of waters flowing in and through the area. The same prohibitions are stipulated in federal law and international treaties that govern protected areas.

Penetrating this wrap of protective legislation, the first feasibility study for Hingol Dam was prepared four years after the area was declared a national park. The study showed that a large dam was indeed physically possible near the estuary at Aghor. Here, the Hingol mountain range enfolds in its arms a large flat basin that lends itself to the building of a reservoir to capture the waters of Hingol River as they flow through the mud-flats and

67

*On the Makran coast, a married woman is often more than a matriarch: her marriage
may be a link between widely dispersed communities.*

estuary to the Arabian Sea. Among only 40 per cent of the world's rivers to be left undammed, the possibility of Hingol Dam must have presented itself as a tantalizing engineering challenge to those who undertook the first feasibility study in 1992.

The dam, if it were ever to be made, would present serious geotechnical design challenges. It is located near the active triple junction of the Indian, Eurasian, and Arabian tectonic plates. This means that as the earth's crust restlessly moves, this area is prone to generating earthquakes. In fact, there are geological fault lines only about 3 kilometres north of the proposed dam site, where currently volcanic mud eruptions are active; those same mud volcanoes that are so revered by the Hindus as part of the Hinglaj pilgrimage. Engineering calculations show, however, that the level of danger presented by this earthquake threat is below what would cause 'a great life hazard'. International guidelines for dam construction allow for this level of danger and factor it into the design specifications and costs of construction.

For engineers, another more problematic aspect also presents itself in conceptualizing a dam at Hingol. The rainfall in this arid catchment area is very low, and data for river discharge is poor. Hydrologists do know that the annual inflows of water from the river are highly variable, so the water resources would have to be tapped to the maximum and a large storage area provided to keep up a regular supply of water. This means that water loss from the reservoir through evaporation would also be a problem. But these factors too can be accommodated into the engineering design and costs of the dam.

The most fundamental enigma to riddle prospects of Hingol Dam remains the question of who the dam would actually benefit. There are about 4,000 people who live inside the Hingol area, and the same number live around its rim. They are mostly traditional semi-nomadic tribes, pastoralist herders, and small fishermen, most of whom practice moderate amounts of rain-fed subsistence farming. If agricultural waters were to be made available to these communities, sociologists know that they are not likely to suddenly switch their livelihoods and life patterns to intensive agriculture. Instead, they will be marginalized and overtaken by land speculators with clients who know more about cultivating cash crops. This displacement of poor peoples is one of the classic themes of public controversy surrounding large dams all over the world.

Instead of claiming local benefits, Hingol Dam was originally conceived as a water sales point for the United Arab Emirates (UAE). In the early 1990s, a meeting was held between the presidents of UAE and Pakistan, where it was agreed that Pakistan would export drinking water to the UAE. At that time, Hingol Dam was designed to export 74 per cent of its water resources to the Emirates. But shortly thereafter, the UAE began to invest in water desalinization plants in order to treat its own plentiful supply of sea water; the world's most cash-intensive way of procuring drinking water was available to them. After this development, Hingol Dam should have been no more than a lame duck, yet it continued to be put forward to a volley of federal ministers as a charismatic project that would win kudos for the government.

The watershed in the Hingol Dam controversy actually coincided with the release of the report of the World Commission of Dams entitled: Dams and Development in 2000. Prepared by an independent commission of experts in partnership with the World Bank and the International Union for Conservation of Nature (IUCN)–World Conservation Union of the IUCN, the report highlights the controversies surrounding the construction of large dams, and in doing so, forever dampened the engineering glory surrounding dams.

At this time, the World Bank was in the process of making a ten million dollar grant to the Government of Pakistan for the Protected Areas Management Project, an environmental project which earmarked Hingol National Park as the most important protected areas site in Pakistan on account of its rich and unique diversity of coastal, estuarine, lowland, and montane habitats that are home to a range of endemic plants and several threatened animal species.

The epitome of survival by substance, the desert nomads create simple shelter from driftwood and desert vegetation.

A Khosa Baloch encampment near the Hingol River is a testament to human and natural coexistence over the millennia.

It was the World Bank itself, once the biggest proponent of dams for development, that expressed its reservations about the proposed Hingol Dam. Unless the government withdrew its dam plans within the national park, the World Bank asserted its reservations about forwarding grant assistance for the conservation project. In writing to senior government officials in 2001, the World Bank stated that any dam in Hingol would seriously compromise the ecological integrity of the protected area, which would breach international protocols on conservation. But in between those lines, the Bank also knew that the World Commission on Dams had admonished it with a classic case. The Hingol Dam has all the controversial elements relating to the questions of what the dam would do to the river flow, uprooting livelihoods and culture of local communities, degrading environmental resources, and offering a poor choice for investment of public funds. As expected, the Government of Balochistan gave an assurance to the World Bank the very next month that the dam would not be constructed, pointing to the fact that the latest policy document on water resource development issued by the government entitled 'Vision 2025' does not include the Hingol Dam.

Yet as the paper trail fades on the Hingol Dam issue in 2002, an internal summary sent to the President of Pakistan leaves behind hints of niggling doubts about the dam. While the correct language is in place in these documents promising integration of custodian communities in park management, concern for integrity of the park, and protection of its innate biodiversity, the note ends with the statement that an 'Environmental Impact Study will be undertaken, ... the Hingol Dam Reservoir Project is the lifeline for the people of the area promising sustainable livelihoods for them ...' before it closes with the statement, 'for your kind consideration to the Chief Executive of Pakistan.'

A highway is built through the Makran wilderness for the first time in its ecological history.

Makran Coastal Highway

Until recently, it took several days of torturous, back-breaking travel through dirt roads to reach the coastal towns of Ormara, Pasni, and Gwadar in Balochistan. For centuries, the 750-kilometres-long Makran coastline had been difficult to traverse. Historians write that Alexander the Great's army almost perished in the harsh terrain of the Makran on its way to Persia. In 2000, the Government of Pakistan embarked on an ambitious plan to build the Makran Coastal Highway, connecting Karachi with the Pakistan–Iran border. The Frontier Works Organization (FWO), the construction department of the Pakistan Army, has built a double-lane road that, once completed, shall be 653 kilometres long. While economists herald the opening of the road, which they feel will boost trade activity at future port towns, environmentalists are concerned about the impact the road will have on a sensitive wildlife habitat. A large portion of the road passes through the Hingol National Park, Pakistan's largest protected area.

It was in 1994 that the federal cabinet gave approval for the construction of a major road through the remote Makran coast. However, work on the coastal highway finally began in 1998. In 2000, the cabinet of General Pervez Musharraf decided to turn the road into a highway with international specifications and to give the mega project the highest priority. A year after President Musharraf conducted the groundbreaking ceremony for the Makran Coastal Highway near Lyari in 2001, the FWO set up a base camp in Aghor at the mouth of the Hingol estuary. The camp was situated in the heart of the National Park, where heavy machinery and drilling equipment for the road were stored.

Construction of the highway was divided into three sections: from Lyari to Ormara (248 kilometres); Ormara to Pasni (197 kilometres); and Pasni to Gabad on the Pakistan–Iran border (208 kilometres). The FWO forecasts that the construction of the road will benefit the isolated fishing communities found on the coast. According to Colonel Hamdani, the commanding officer at Aghor, the highway will directly benefit the people living in this region.

Cut off from the outside world for centuries, small fishing towns like Kund Malir will now have access through the highway to better healthcare and schools in Karachi. Before the construction of this road, the people of Kund Malir had no access to medical care and the only school in the area was a boys' primary. There was also an acute shortage of clean drinking water in the settlement.

With the completion of the highway will come electricity, phone lines, and water tankers. The road will also bring in new construction, commercialization, and settlers from elsewhere. New settlers are bound to introduce new diseases and different lifestyles to a local population living in what is one of the world's most sparsely populated locations. The traditional hierarchy of the small village could be changed forever as more powerful groups of individuals move in and buy land. Already, land speculators are quickly buying up land from the fisher folk all along the coast for future resorts and housing estates. Along the Makran coast, land prices went up from six thousand rupees per acre to more than one hundred thousand rupees per acre in just one year in 2001. After selling their land, these fisher folk are moving to big cities like Karachi where they face a host of problems and end up living in squatter settlements.

In Hingol itself, the road has been built without any environmental review and minimal standards for determining location. No environmental impact assessment had been conducted prior to the road's construction. Developers will no doubt be initiating new construction alongside the road inside the National Park as well, unless legislation is passed to protect this fragile ecology. Hingol needs to receive permanent protection as a wilderness area that must not be disturbed any further. The road will interrupt the movement of wildlife in the park, which was rated by IUCN as among the significant protected areas on earth in 1997. Animals are habituated to moving around freely in what has hitherto been a natural system.

*A local woodcutter walks the route that nearly destroyed
Alexander of Macedon's army in 325 BCE.*

The asphalt road itself has been designed to have a 7.3-metre-wide carriageway with 3-metre-wide shoulders and fences around it. Since the Lyari to Ormara section is infamous for its flash floods, a large number of drainage structures have been provided. This includes thirty bridges, seven hundred and six culverts of varying sizes, and six hundred and fifty causeways. The longest bridge of the highway is over the Hingol River, which will be 400 metres long and 8 metres high. Flash floods caused by hill torrents are common in this region and the builders of the Makran Coastal Highway, which is being developed along the same route that Alexander took to Persia, are well aware of this problem. Flash floods, of course, almost destroyed Alexander the Great's army during his last, terrible march to Persia in 325 BCE. Historians say that his army was caught in a flash flood near Hingol. Most of the camp followers were drowned—Alexander and his troops barely escaped with their weapons. In July 2003, an unprecedented spell of heavy rains damaged a large portion of the completed highway between 3 and 24 kilometres long on the Lyari to Ormara section.

The Gwadar to Pasni section has already been completed with almost the same specifications as the first section. It has 15 bridges and 421 culverts. The FWO's accelerated schedule of work has enabled the two stretches of the highway from Lyari and Gwadar to meet in 2005. The ports of Ormara, Pasni, Gwadar, and Jiwani are all located along the coastal belt; it is hoped that the completion of the highway will boost Pakistan's multi-million dollar fishing industry as the fisher folk living along the coast will now have easier access to the bigger markets in Karachi. At present, they are completely dependent on the middlemen who have a monopoly in the area owing to the expenses of transporting fish to Karachi. The middlemen buy their fish and then sell it in Karachi at a much higher price, cutting the fishermen out of the profits.

The coastal highway was opened to the public in 2005 with a government expenditure of Rs. 15 billion. While there is no doubt that Balochistan's promised riches of marine resources will be fully exploited for commercial purposes, it remains to be asked at what price to the public exchequer and for whose benefit these will be used. The following essay illustrates the disturbances that a new highway through the wilderness causes to a previously undisturbed ecosystem.

The coastal highway was opened to the public in 2005, but at what price to the public exchequer and for whose benefit?

The reptiles of Hingol's desert are admirably equipped to deal with the scorching sun,
but cannot cope with the superheated surface of the coastal highway.

The Coastal Highway: Simple, yet Sinister

One hot summer's night in 1947, the Maharajah of Korwai, in what is now northern Madhya Pradesh, shot to death three cheetahs as they stood transfixed in the headlights of his official car. Recounting the incident in his book, *The Mammals of Pakistan*, Dr Tom Roberts wryly recorded that, 'there have been no authentic sightings [of Cheetah] in India since that date. ...'

Just a half century later and 1,200 kilometres to the west, a project was inaugurated that would introduce all of the key components of this tragic episode to the fastness of Pakistan's Makran coast: people, cars, lights, guns, and avarice. The project was to construct a coastal highway linking the hitherto relatively isolated towns along the desert coastline, from Karachi to Gwadar.

Hingol's shifting sands provide a brief and often artistic record of what has passed an hour or even a night ago.

Roads are so often revered as symbols of human development that the term is pervasive in English metaphors for progress. But they are so often also the harbingers of environmental destruction. What will the ecological and sociological price of this coastal highway be? In all probability, no one will ever know, for the development of the road has preceded the introduction of an effective management programme to Hingol National Park and ended the isolation of the thinly scattered villages in the arid hills and along the desert shoreline.

It seems inevitable that some indigenous species will be locally extinct before science has recorded their very existence. The idiosyncratic cultural adaptations of many of the Makrani villages, so long the instruments of human survival in this harsh desert landscape, will fade away like starlight in the desert dawn.

The desert knows no natural analogue for a modern, elevated, hard-topped highway; nothing in nature presents a barrier at once so sinister, simple, and complex. Transverse to the natural shoreward watersheds, the road cuts off smaller drainage lines. These force small drainage lines to coalesce into larger streams, giving an unnatural boost to the erosive power of flash floods that emerge seasonally from the hills.

Tracts of dunes are either synthetically stopped or started by the intrusion of a static, immobile component into an environment of moving sand. The black pitch that is used to seal the surface absorbs heat like a lava flow. In the summertime, slow moving reptiles like saw-toothed vipers and desert tortoises are literally roasted before they can cross the burning tarmac — the road becomes a continuous thermal barrier.

*In Hindu mythology, Chandragup, the large mud volcano of Hingol, is believed to be the head of Shiva,
the god of destruction. Pilgrims throw coconuts into the crater at the top to appease the god.*

Photographs courtesy: Hasan Zaki by permission of WWF-Pakistan and PWP

In the wilderness that is Hingol, among the semi-marginalized
people who inhabit it, lives a pastoral chief called Talal.

Talal

Despite the burgeoning human population of the flood-plains that flank its mighty rivers, Pakistan still has great tracts of wilderness; broad, seamless landscapes mostly unmarred by the marks of humanity. Callously characterized and condemned as barren ground, there is great beauty in these wild places. Wilderness, especially deserts, highlights all life that exists in it. Even a lonely daisy at the base of a tall sand-dune shines jewel-like in the sun, and an eagle high on a windswept crag seems an icon.

Wilderness ebbs and flows with the tides, fades with the daylight, and flourishes in the dawn. Seasonal changes soften its hard countenance, and when they do, there is a startling array of life forms that capitalize on the brief respite that nature presents: grasses and wild flowers that emerge from nowhere, and brittle, apparently lifeless little shrubs that flame into leaf. The animal kingdom profits from these temporary times of plenty and migrate with them, gently, like the shadow of a cloud on the ground. The collective urge and need for such migration have almost faded from humanity, but in the deserts of Pakistan, there are still a few scattered groups of people who live in and with the wilderness. They earn suspicion and animosity for doing something that comes completely naturally, following the green grass.

In the wilderness that is Hingol lives a pastoral chief called Talal. He is a tall, wiry man with craggy features and deep-set, unblinking grey-blue eyes. A clean scimitar scar runs diagonally across his left cheek like a crack in an antique vase. Perhaps he was born with it. He wears a flowing robe of light brown homespun material and a matching turban with a long tailpiece that drapes over his shoulders, framing his bearded face. His hands and forearms are hard and sinewy, like a weather-beaten juniper.

To a visitor, his physical image frames the fortitude of his character. How resourceful and resilient he and his people must be to live such a lifestyle. He is a tough, hard man, whose existence has been a struggle, but ironically his name is Talal, Arabic for 'dewdrop'.

Introduction

Sonmiani is a small, historical community on the Makran coast of Pakistan only 90 kilometres west of the metropolis of Karachi. Little known to the nation, it is one of many communities that go about their traditional ways of life in their own quiet and peaceful ways. Fishermen by trade, their lives are interconnected with the natural environment of there habitat. The bay is home to Balochistan's major mangrove forests that provide an abundance of fish for trade and food, as well as fodder and fuel for the local residents.

Yet in this seemingly stable existence, a number of problems threaten the livelihoods of the people who live there. Sonmiani, like many other local communities in Pakistan, is facing a crisis that has at least three major sources. First, locals are not aware that the natural resources—in this case, the fish and the forests—are not unlimited and will not last forever. Second, subtle changes in the rivers that flow into the bay from the mainland are causing the bay to silt up and change its natural characteristics. Finally, middlemen keep accumulating the debts that people owe them so that financial freedom is unachievable for many.

In the face of poor government services and weak application of law, the fishermen of Sonmiani have little security for future generations. This is the story of the ways in which solutions can be brought about from within the community, and of the human efforts that may change the course of many of these impending problems.

The story also shows how outside agencies can play a critical role in bringing about local solutions that are waiting to happen. Like the process of human birth itself, in which the presence of a midwife can be decisive, an outside agency can facilitate solutions from within the community. But like a good midwife, the agency must be attentive, resourceful, and equipped with scientific knowledge.

In the five years of effort, two out of three communities of seven thousand people have organized themselves to act together, and incrementally solve many of the problems that beset their lives.

(facing page)
Fishermen play a form of chequers resembling chaupan, an ancient subcontinental game thought to have originated during the Harappan civilization.

Forests have been revived, girls educated, sceptics won over, and local technologies developed. Slowly, neighbourhood boys will become village leaders and these same men shall rise to become regional forces.

Unfortunately, though the immediate problems have been addressed, long term and complex realities still need to be tackled.

We observe how this empowering change happened and who brought it about. The nature of this change gives us much to learn about how people survive adversity, and continue to live their lives within their means, and with dignity. It has lessons for us as citizens as well as for governments and those who make the policies and laws of the land.

The story of Sonmiani is just one piece of a new pattern that is emerging in Pakistan; a pattern that redefines the way that local communities are taking charge of their own lives and forming new relationships with institutions such as local governments and elected representatives, that form part of the fabric of a changing society.

Greater flamingos wade in the mangrove creeks where shallow saline water affords them food: on small shrimps, seeds, algae, and mollusks that they suck and filter through their bills. The distinctive pink coloration of the plumage comes from pigments in the organisms upon which the flamingos feed.

Photograph courtesy: Ghulam Rasool by permission WWF-Pakistan/PWP

From the day they are born, the sea becomes vital for Sonmiani's inhabitants.
Boys prepare for a morning fishing expedition.

A History of Sonmiani Bay

Sonmiani Bay is situated at a distance of 90 kilometres from the metropolis of Karachi. Located on the eastern edge of the Balochistan coast, it constitutes the smallest *tehsil* (administrative unit) of the district of Lasbela. The *tehsil* comprises three small villages, including Sonmiani, Damb and Bhera. The population is estimated at about 7,000 people, of which 4,000 live in Damb, 2,000 in Sonmiani and 1,000 in Bhera. Today a little known *tehsil* of Pakistan, the bay area has undergone many dramatic changes in the past as it was transformed from an important port to a small fishing settlement.

Records from the British Empire dating back some 200 years identify Sonmiani Bay as one of the more important harbour and trade areas that provided access to Sindh, as well as the Indus River trade routes. British explorers carefully searched for viable harbours along the Makran coast, and researched the ancient trading ports that had been used by the early Muslim Arabs in the seventh century CE as well as during the preceding Greek era of Alexander of Macedon whose reign lasted from 332–323 BCE. Lieutenant Henry Pottinger, among the earliest of the British travellers who preceded the formal British presence in the area in the mid-1800s, writes in 1816:

> The Bay of Sonmeany ... is formed by Cape Monze and Chilney Island on the one side, and Cape Urboo on the other: it is a very noble sheet of water, said to be free from rocks and shoals, and is capable of affording anchorage to the largest fleet: it is celebrated as the rendezvous of that of Nearchus ... the description given ... of the Port of Alexander so exactly corresponds with its actual state. ...

In addition to claiming that Sonmiani was in fact the port at which Nearchus, one of Alexander's generals, led a contingent of the Greek army that was to set sail to Persia, Pottinger also describes the goods that were traded from the hinterlands of Balochistan and Sindh. The exports that were shipped to Arabia

Sonmiani, today a little known tehsil of Pakistan, was once visited by Alexander of Macedon, Marco Polo, and the adventurous informers of the East India Company.

Indo-Bactrian coin showing image of King Diodotus (2nd BC) from the central Punjab region. Image courtesy, Keeper of Coins, Lahore Museum.

from the bay were mostly grains, some felts, and coarse carpets. In exchange were received dates, almonds, and 'Caffre slaves' about whom he emphasizes that, 'they are very valuable and (did) most of the outdoor labour.' From Bombay, imports through Sonmiani consisted of iron, steel, tin, sugar, betel nuts, and cocoa nuts. In exchange, Sindh would send coarse white cloth, chintzes, 'loongees' or the cotton sarongs still worn in rural areas, and a little raw cotton intended to be worked into a stuff called 'Khargee'.

From historical records, we can safely assume that the inhabitants of Sonmiani were engaged in trade of an important and vigorous nature, supplying the incoming merchants with goods and receiving products of considerable value. The trade in slaves from Arabia and Africa is likely to have been included in the markets of Sonmiani. In CE 1289, Marco Polo records about the coast and its people: 'they live by trade and industry ... they have rice, meat, and milk. Merchants come here in great numbers by sea and by land with a variety of merchandise and export the products of the kingdom. There is nothing else worthy of note.'

Two centuries ago, Sonmiani was a *riasat* (principality) of Lasbela, belonging to the hereditary tribal ruler or Jam. In the vicinity of what is now the local boys' school stood the Jam's palace, where taxes were levied on all the commodities coming into the bay. This practice was still in place after the partition of the subcontinent and the creation of Pakistan in 1947. Lasbela remained the property of the Jam until succession was attained as part of the unified State of Pakistan in 1957. As with other principalities and states, special concessions were given to the Jam by the government. To ensure that his traditional leadership continued, he was accorded political representation via national and provincial seats in parliament reserved for him and members of his family. In matters of politics, the local people of Sonmiani are still affiliated with the Jam, their traditional leader.

*The dai or midwife of Sonmiani, whose experience is supplemented
with training that certifies her as a traditional birth attendant.*

People and Ethnic Origins

Today, a visitor to Sonmiani would be astonished to learn how the geographical location of the bay has changed over the past centuries. In Sonmiani village, derelict mud structures that were once shops stand outside the mosque. This is the original bazaar of the historical bay, close to where the trading boats of Arabia and Bombay (Mumbai) came into harbour in the deeper waters of the bay. Dominated by Hindu shopkeepers, Muslims would participate in the activities that supported the trade; for example, they worked as hired labour in godowns, but none of them owned property.

Pottinger reports from his Sonmiani visit of 1816: 'I was quite astonished to find so much trade going on ... the commerce is entirely monopolized by the Hindoos, whose indefatigable industry is conspicuous where ever they are to be met with...'

After supporting the trade of the bay for centuries, a hiatus arose in the social fabric of Sonmiani, when the Hindus left at Partition. Around this time, nature also changed its course. The Winder River, which was a major tributary bringing fresh water into the bay, began to silt up. This meant that the waters of the bay became shallower and less tenable as a harbour for trade boats. The inhabitants of the once prosperous market began to move west into the bay and reside in an adjoining hamlet, a better harbour that came to be known as Damb.

At this time, when the Hindu inhabitants fled and the inflow of one source of fresh water dried up, new settlers came to Sonmiani. These people, whose descendants are the present inhabitants, were settlers from Kalmat Hor and Hingol, upper Lasbela, and Sindh; areas of scarce natural resources and harsh living conditions. Some of the original Muslim inhabitants of pre-Partition Sonmiani also remained and have come to play an important role in the area. But it is worth noting that most of the people who presently live in the bay are in fact settlers who arrived only about fifty years ago, spanning a mere two generations.

A few major clans inhabit the bay today—the largest one is the Khaskheli clan. They have African features, and were once servants

The smiling hopeful youth of Sonmiani. Little boys play a large role in running errands for clan neighbourhoods.

of the Jam, who paid them in kind. After Partition, these people began to educate themselves. The Jumari Baloch claim to have come from Kalmat, the Somar Baloch or Soomro from Sindh, and the Zikri Baloch say they have migrated from Kalmat and Hingol.

The Hindus who remained in Sonmiani after Partition were those who were not able to flee the beleaguered subcontinent. They were given protection by the Jam to stay on. Presently, there are about a dozen Hindu households in Sonmiani. What is particularly noteworthy is the trend amongst the different clans and faiths in Sonmiani to live together in harmony. In 1992, when Babri Mosque in India was destroyed by right-wing Hindus, people came from all over Winder to destroy the *mandirs* (Hindu temples) of Sonmiani in retaliation. However, the young boys of the area encircled the *mandirs* to protect them from these outsiders. They felt that the Hindus of Sonmiani had no hand in the events in India and that they had always dealt with their neighbours fairly.

Neither do the local clans enter into conflict, respecting each other's autonomy. There is etiquette involved in how young men interact between clans. People send little boys to deliver messages to members of another clan while they themselves remain outside the clan neighbourhood. Meetings take place in the *autaq* or men's area of the clan neighbourhood, as a mark of deference to the women of the clan. Although they remain segregated and do not participate in public or economic activities, women are accorded respect. No murder or theft cases have been reported in Sonmiani in the last five years, primarily because of the vigilance of this close-knit community.

*We now understand that mangroves have the power to buffer the devastating
effects of shock waves of water that destroy life and homes, after
witnessing hurricanes and tsunami hit the shores.*

Ecological History

Sonmiani Bay is a closed lagoon with a four-kilometre-wide mouth which opens out into the Arabian Sea, and through this feature nurtures life forms that are distinct from those of the open shores of the Makran coast. Much of the lagoon waters lie within convoluted land formations and islands. The lagoon is about 60 kilometres long and its width varies from 7 to 10 kilometres. Vital to the life forms that are found in the bay are the two bodies of freshwater that flow into the bay from inland: the Porali River drains through the Bela region and the Winder River drains near the mouth of the bay. Standing on the shore of Sonmiani Bay, one sees a wide, open body of murky water with dark green vegetation on the horizon. Once on a boat sailing into the waters, there are Bottlenose dolphins, turtles, and a multitude of water and sea birds to be seen.

The flow of water in the Porali and Winder Rivers, however, is decreasing—a part of this can be ascribed to the construction of the Hub Dam, which provides drinking water to the city of Karachi. This means that the hydrology of Sonmiani Bay will be severely affected as time goes on. One of the key ways of maintaining the Bay as a productive wetland is the inflow and outflow balance of fresh water. Any change in this delicate balance means that both the physical and chemical characteristics of the bay will be altered. It is difficult for scientists to project what might happen in the future, but the three species of mangrove trees found in the bay require a specific water salinity to flourish. These, we know, are nurseries for a large part of the fish population of the Makran coast, where eggs are laid and hatched, and young fish grow before entering the deep waters of the ocean. These subtle and slow alterations present a looming biological problem for the future of Sonmiani Bay. Another side effect is the gradual drying up of drinking water wells, on which the people living near the estuary were dependent. This is causing people to migrate from the area.

Before work on environmental conservation started in the area in the early 1990s, the villagers thought that the mangrove forest was a natural phenomenon that should be left on its own and used

In the mid-1990s, Sonmiani's fishermen began to notice that their harvests of fish and shrimp from the bay had declined quite dramatically. This motivated collective action for the conservation of the mangrove forest and the life forms that it supports.

This is the fluid landscape of the fishing communities who make this bay the source of their earnings and daily labours.

so that it regenerated on its own. As a consequence of overuse and neglect in the past ten years, shrimps and fish had declined dramatically in the bay.

The large trawlers that ply the high seas outside the bay and harvest fish in the most intensive way are owned by fishing companies holding government permits. They employ seasonal workers from Burma, Bangladesh, and the Makran, all of whom flock to the bay as a living base during their time on the coast. The poorest of the local people are most aware of the eroding fish catch because it has a direct impact on their income. But the *waderas* or wealthy fishermen are unaffected by this environmental decline, since the large size of their catch makes up for the diminishing volume of fish in the longer term. They remain unaware of the destruction they are causing, and yet this will be a crisis for everyone in the long term, rich and poor alike.

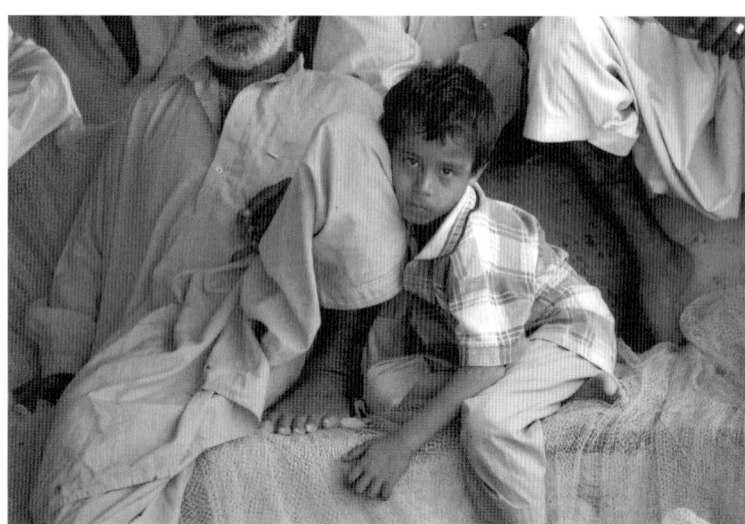

Young and old alike gather to hear the discussions that affect the lives of fishermen throughout the coastal areas of Pakistan.

*The repair and maintenance of fishing nets is a skill essential
for the small- and medium-scale fishermen of the bay.*

Men play a friendly game of cards during the seasonal lull in fishing activities.

The Power of the Middlemen

A nother problem that the people of Sonmiani Bay are facing today is increasing debt. The Karachi-based businessman, usually from the canning or fishing industry in Karachi harbour, appoints a man in the area to work for him. This middleman invests in trade and pays for the repair and maintenance of the boat and nets (one of the most expensive assets of a fisherman). He also underwrites the off-season expenses of the fishermen at the teahouse. Then the fisherman becomes bound to sell his fish to the middleman. If the market price of the fish is four rupees, the middleman will pay two or three rupees. In recent years, some agitation has taken place to negotiate better terms from the middlemen. The concessions the middlemen make to the local fishermen allow them to sell a portion of their catch in the open market. But the Karachi businessman who earns the biggest profits from the industry in Sonmiani does not give anything back to the Bay.

Khushali Bank, a microcredit banking facility offered by the government, also provides loans to small groups of fishermen who then appoint their own middlemen. But collateral is needed to bank and savings are essential. However, the concept of savings is not common practice in this community, as barter and loans have prevailed in its local economy. When the cash flows into

A Sonmiani schoolteacher with influence over local youth, Master Rashid regularly organizes group sports for social change.

Interest in community participation has led young activists like Jamil, from Damb,
to be elected Naib Nazim or Deputy Mayor of Lasbela.

households after the auction of fish has taken place, the women joke that the wheat of their area does not taste good enough for them. The shopping expeditions of Sonmiani's women to Karachi after the fish harvest are the bane of their husbands, and teahouse jokes relate how women even buy their *lotas* (jugs used for ablution) and wheat (a daily staple) from the big city.

Today, the ideals of well-being are also changing in the bay. New concrete and tiled houses owned by shrimp farmers and middlemen are setting new trends for housing construction. Now, many of the villagers want these houses instead of the traditional mud plaster and palm matting ones. To attain this way of life, households in Sonmiani will want more access to cash and loans. As a reflection of these new social pressures, fishing regimes are increasingly intensive. Whereas an average fishermen would fish for in the bay three or four hours, he may now be willing to fish for up to fifteen hours.

The people of Bhera decided to mobilize themselves without outside intervention when they realized the benefits that were being enjoyed by their neighbouring villages.

The Arrival of WWF

Outside agencies like World Wide Fund for Nature (WWF) in Pakistan have played an important role in helping the people of Sonmiani find solutions to the problems facing their communities. In the early 1990s, WWF started work in the bay by trying to understand the biological and social problems of the area, focusing especially on finding local solutions for improving mangrove forests as well as developing forest rehabilitation techniques.

The scientific work carried out by WWF was an essential element of their presence in Sonmiani. In tandem with the work with community organizations, WWF started learning about how the three species of mangrove plants available in the forest of the bay actually grow. Studies were conducted to learn about the rehabilitation of the forest by singling out the most important pressures on the forest. A first attempt was to grow a nursery of the forest species along scientific lines, and observe progress of the species using growth charts that measure number of leaves, spaces between nodes, branching, and other indicators of plant physiology. Salinity and pH levels were carefully measured. They found that two species, *Rhizophora mucronata* and *Ceriops tagal*, were easy to grow in nurseries, but *Avicennia marina* showed little growth in these conditions. The seed germination of this plant had only a 10 per cent success rate.

The successful approaches that WWF discovered through practical experience were based on mimicking nature as it occurs in this particular wetland. They found that a seedling nursery must be located in a place that receives natural irrigation from the ocean tides. Watering should occur during high tide, so that the seedlings are neither flooded nor dry. The most delicate of the three mangrove species, *Avicennia marina*, only responded to transplantation at a certain time in its seedling growth. Its complete life cycle had to be understood before its transplantation into the forest could be mastered. WWF called this technique the transplantation of wildlings. Careful scientific methods were used to monitor the best places to transplant wildlings. WWF collaborated with Karachi University's Marine Biology Institute to develop the scientific understanding to best ensure survival.

The science involved in understanding the nature of mangrove forests has been essential to the conservation work that communities are now doing. The mangrove forest is the basis of the abundance of fishery that is possible in Sonmiani. This industry is the mainstay of Sonmiani's economy today. As local communities have realized the importance of saving and regenerating the mangrove forest in order to keep their own lives tenable, they have looked to the scientific work done by WWF as necessary for mangrove conservation, for they are the primary users and beneficiaries of the natural gifts of the mangrove forests.

*Public discussions and meetings with outsiders ensure ongoing reflection
on the social and economic issues of the bay.*

Coastal Communities Come Together

In 1995, a Fisherfolk Forum was founded in Karachi to protect the interests of local fishermen with the help of Shirkat Gah, an NGO working for women's empowerment, and the International Union for Conservation of Nature (IUCN). The chairman of this forum, Mohammed Ali Shah, remains actively focused on coastal fishermen's rights for Pakistan. He initially approached the WWF to assist in reaching the Sonmiani community organizations and was invited to the sports event in the bay area in the 1990s. Later, in 2000, the Community-Based Organizations (CBOs) at Sonmiani were invited to join the annual Fisherfolk Conference. They eventually became partners of the Fisherfolk Forum and used this coast-wide association to share their successes and exercise influence in government and political arenas.

The fishermen of the coastal areas are passionate about football, especially during the off season when they are not engaged in fishing activities. Traditionally, the influential people of the area would sponsor a district-level tournament every year. On one occasion, a conflict arose between Sonmiani and Damb in 1994, caused by a foul that took place during a football game; this caused a fist fight. Although the quarrel was resolved by the elders of the villages, the tournaments ceased to take place after this incident. However, people yearned for sports events, and the WWF decided to revive this local tradition by adding an awareness campaign to the tournament as a special feature.

A mangrove plantation campaign was planned along with the first revived tournament in 1995 as a test case for the CBO in Sonmiani. The community spirit during this event was excellent and people organized the event independently. During the game intervals, the commentator incorporated various facts about mangrove forests. The first prize was given out by a shrimp trader, who spoke about the importance of the relationship between mangroves and shrimp fishing. So the passion for football became an innovative forum for bringing the importance of mangrove forests to the centre stage of community events.

After this enthusiastic response, the WWF organized other events with the Damb CBO at the *tehsil* level. There were regular events every year and the games diversified to cricket and other sports. Eventually, the CBO began to take over all the organization and fundraising from the big companies in Winder, while the WWF became a guide for composing the technical messages and the prize-giving ceremony.

In this coming together of diverse groups, we see the initial weaving of a new social fabric that rose organically from within the local milieu. This is how new social institutions are created where there is a hiatus caused by the disintegration of historical structures in society.

Damb Village: The Pioneers

Amidst the bustle of Damb's fish marketplace are the seeds of leadership that first manifested themselves in 1997. The centre of economic activity in the bay, Damb is open to the outside influences that come and go with its trade with Karachi. The trappings of urban culture come in many forms and amongst other things, the people of Damb began to notice an upsurge of 'mini cinemas' and video movie houses showing vulgar entertainment. In this traditional rural society, many eyebrows were raised, and a group of young men decided take action to curb the new influence that so offended local sensibilities. They approached the owners of these cinemas and negotiated a settlement to put a stop to films they considered objectionable.

Their success with this small action made the group realize that they had been effective in exerting their influence to bring about some change for the better in their community. As a result, an *Anjuman* or youth development organization was formed. This organization set up an eye treatment camp, networking with doctors in Karachi to provide free services to villagers as well as volunteering their own services. At this time, in 1996, two officers of the WWF came into Sonmiani to assess the possibility for starting a community-based mangrove conservation project.

They recognized the young men's networking abilities, but observed that this organization only came together for some issue to which they wished to react and then dispersed. Through its professional community development officer, the WWF helped the nascent CBO to establish a mandate and regularize office bearers, rules, and regulations. The WWF then began to work on building their skills in various management aspects; for example, the role of the activist, the management of the CBO, and the keeping of accounts and auditing.

With a group of seven young men in their early twenties, the small organization began to define various divisions of its work, including that in the health, water, education, and environment fields. Discussions with WWF revealed that the locals viewed the natural resources, especially the mangrove forests, as undiminishing. But when they were made aware about global problems in forest resources, it was easy for them to understand the need for conservation.

By 1998, the work on community mobilization had matured with the capacity of the CBO to manage its accounts, administration, and social mobilization. At this time, the cost and techniques for mangrove rehabilitation were also ready. The WWF began to introduce these techniques as extension skills into the CBO and also organized trainings on mangrove nursery raising. Slowly, the CBO came into more advanced levels and its members began to develop their vision into a larger perspective.

As a result of this evolution of action coupled with social responsibility, the leading young man of the Damb CBO, Jamil, was elected as the *Naib Nazim*, or deputy mayor, of Sonmiani. The development perspectives of the residents of Sonmiani Bay were now actively represented through him in the elected council. The Damb CBO played a catalytic role in embracing new approaches in Sonmiani Bay.

Sonmiani: The Young Institution

After being approached by the WWF, it took Sonmiani village four years to form a CBO, as they slowly weighed the options for and against such a move. The people of Sonmiani are known to take a long view of new interventions and are more conservative in their outlook. Unlike Damb, where society is much more open, and locals exemplify this openness by saying that 'love marriages' are socially acceptable, marriages in Sonmiani are predominantly arranged by the elders, and those who marry against their family's wishes are ostracized.

In the nascent Damb CBO, the vice president was a resident of neighbouring Sonmiani village. For the first two years of community mobilization, Sonmiani residents were part of the Damb CBO. They observed the way the Damb CBO grew and tried to understand the new dynamics being created in the community. Traditionally, Sonmiani had three small *mohalla* (neighbourhood) based youth organizations. These little groups would meet on the last or first Friday of every month and collect money to assist in marriage or funeral arrangements. They did not think much further than these functions. The WWF was invited to introduce their work in the village and they suggested that Sonmiani should develop a joint organization of all the *mohalla* groups.

In the cautious society of Sonmiani, the WWF had to prove its worth to the residents of the village despite of their achievements in nearby Damb village. The pre-eminent needs identified by the people of Sonmiani village were health problems, related to reproduction and childbearing, faced by the women of the village. The WWF won the confidence of the people when they responded to this need by starting a midwifery training programme. Although this falls beyond the mandate of mangrove conservation, both WWF and the people of Sonmiani were able to see that social development and conservation must go hand in hand.

After much deliberation, the Sonmiani CBO took shape and called itself the Sonmiani Development Organization. They elected a Hindu from the village as their vice chief patron, displaying the respect that they accord to this minority group. A culture of tolerance and careful deliberation is the hallmark of this CBO, which has carried the collective efforts of the community into a mature and holistic organization for the aid of community needs. Recently, the CBO was awarded an international forest conservation grant award on behalf of all the community organizations of Sonmiani Bay.

Bhera: The Sceptics

Bhera is an isolated settlement whose people are Zikri Baloch, a little known Islamic sect confined to Makran and Turbat. The Zikri sect was founded about five hundred years ago in the Indian province of Gujarat by Sayyid Muhammad Jaunpuri (CE 1443–1505), who proclaimed himself to be the promised Mahdi (spiritual and temporal redeemer). He won many adherents, including members of the Mughal court, but made many more enemies, especially amongst the *ulema*. When he died, his spiritual mantle was taken up by his son, who was incarcerated by the state and died in prison. By 1576, the movement had virtually collapsed and its followers dispersed. Nevertheless, there are still small communities in Iran, the Gulf, India, and in Pakistan's Balochistan and Sindh provinces.

Until the 1980s, the Zikris were peacefully absorbed into the State of Pakistan. When the country's orthodox *ulema* succeeded in getting another small Islamic sect — the Ahmadiyyah — declared constitutional heretics and hence 'non-Muslim', they began a similar campaign against the Zikris. By the late 1980s, the Zikris were being actively persecuted. This led to the further isolation of these people in the Makran coast. They are further distanced by the fact that they belong to another linguistic group. The people of Damb and Sonmiani claim to speak 'Lasi', which is in fact a dialect of Sindhi. In Bhera, they speak Balochi and were originally migrants native to the Hingol area.

The poor road connection to Bhera is another major factor in cutting off these people from the outside world. But even with improved roads, the community of Bhera will remain conservative and wary of the outside world, in sharp contrast to the extroverted Damb community. The few scholars who have studied the Zikris believe that they are amongst the spectrum of Sufi or mystical traditions that have evolved in the Islamic world. There is great emphasis on *zikr* (remembrance) as a form of worship which is a common practice in Islam. The Zikris place emphasis on the equal distribution of wealth and material possessions. Each community constitutes a *daira* or circle and all wealth is collectively owned. The Zikri view of society is egalitarian and denounces personal property with emphasis on mutual help. In contrast to their neighbouring villages, Zikri women participate in the community's public and socio-religious life.

Presently, Bhera is at a crossroads of change. As the more conservative elders pass on, the educated younger generation perceives benefits in the new ways that Damb and Sonmiani village are embracing. The people of Bhera started working with the Sonmiani CBO in 2003, which has proved to be a good strategy. While the elders in Bhera are still suspicious of outsiders and feel that letting new ways and people into the community may allow others to lay claims on their village and forest, the young men of the area have consented to start a community-based organization in their village, eventually making their own CBO.

Today, they have accepted the presence of the WWF in the area and joined in their trainings on fuel-efficient stoves. It is important for the local villagers to use these stoves, which will reduce their demand for fuelwood and thus ease the pressure on the nearby mangrove forests. In 2004, the Bhera CBO began mangrove plantation and established their own nursery. The people of Bhera decided to mobilize themselves without outside intervention when they realized the benefits that were being enjoyed by their neighbouring villages.

The story of Sonmiani is one piece of a new pattern that is emerging in Pakistan; a pattern that redefines the way that local communities are taking charge of their own lives and forming new relationships with traditional partners such as the local government and elected representatives.

Mangrove forests provide the inhabitants of Sonmiani with fuelwood and fodder for livestock.
They are also sanctuaries for fish and shrimp to nest in.

His name is Khuda Dad, meaning 'divine endowment'. He is one of the only two boys in Bhera who have managed to attend college.

Khuda Dad of Bhera Village

Only when the sea tide recedes in Sonmiani Bay does the dirt track to Bhera village emerge. The village is tucked into the western-most crook of a bend in the enclosed bay that is surrounded by mangroves and creeks flowing into the Arabian Sea. A collection of thatched palm huts surrounded by thorny fences and a lone white-washed mosque are all that are visible. A desolate place of two hundred and sixty-five households, the village seems to belong to another century.

There is one teashop where signs of life stir; a group of children hover here to buy sweets and marbles. These children are the only residents who are willing to talk to outsiders, even though none of them seem to speak Urdu. In fact, there is a language barrier with their neighbouring village as well, since the residents of Bhera speak only Balochi while the people in the rest of the Bay speak Sindhi and Lasi. The teashop is half empty, and the two men buying their afternoon *'paans'* shy away from the inside of the clapboard shop that is constructed from the planks of wooden packing cases.

The arrival of outsiders has caused a stir in the village, and a young boy emerges, with earnest eyes and sober bearing. He comes to engage with the visitors as he speaks both Urdu and English. He is intensely interested in the visitors and eager to converse. Perhaps this is the one chance he will get in many months to speak in English.

His name is Khuda Dad and he is one of the only two boys in Bhera who have managed to make it to college. With his spare frame and thin face, he looks about fourteen, but tells us that he is studying his first year at Muslim Public College in Hub Chowki. This makes him at least seventeen years old.

'I want to be an engineer,' he tells us his large intense eyes glowing with determination. In a village without a sign of steel or concrete, this is indeed a striking ambition. There is not one television set or electrical connection in the village. The elders of the village are in no hurry to herald the changes that are sweeping through the outside world. In this tiny community, they feel they are one nation, intact and living in security without crime and conflict.

Born in Bhera, Khuda Dad has struggled through a brave journey to reach this level of education. There is a primary school with one teacher in this village, and most children stop studying at this level. There are eighteen girls in the primary school. For years of secondary schooling, Khuda Dad has had to traverse the diurnal sea tide to cross his village to Damb Bandar where the secondary boys' school is located. Beyond this, the collective efforts of his extended family have been instrumental in finding him a place to live as he pursues further studies, initially at the boys' high school in Winder and then at a college in Hub.

'The elders of my village are suspicious of the outside world and they say that mingling with outsiders creates trouble in our community,' he tells us. Bhera is a Zikri Baloch village and the people are migrants from Kalmat Hor.

Khuda Dad's five brothers are all fishermen and he says they are well aware of the links between healthy mangroves and increased shrimp catches. But he says that they are uncomfortable with the idea of working with outside agencies because the 'elders wouldn't like it'. The elders have only recently agreed to form a community-based organization—traditionally, they have disapproved of such new-fangled ideas and felt that there was no need for this kind of activity.

There are also fears that outsiders might impose taxes on the use of mangrove wood and fish from the bay. They recall that in the 1950s, the Balochistan Government suddenly placed temporary, but painful tax on the cutting of mangrove wood. Most of the villagers of Bhera are camel herders and they still graze their camels in the surrounding mangrove forest.

With a faint parting smile, Khuda Dad leaves the teahouse to return to his home, where he is spending his spring holidays away from college. As he leaves, he says he would really like it if his village could get computers. They might not have electricity or roads, but for the new generation, computers remain a potent symbol of the ideals of the modern world.

RANGLA

CHOLISTAN

LUNGH

SECTION TWO

The Deserts

Introduction

A historical legacy of Shikar, or hunting, during the Mughal era associated a culture of regal masculinity with Lungh Lake; this tradition continued in the form of colonial imperialist character-building during the British Raj. Later, after Partition, it was given the status of a nature lover's bird sanctuary, assiduously protected by the civil service. This history shows how a group of committed nature lovers and scientists laid the foundation for bird conservation for decades beyond their active lives, discussing also the conflicts inherent in a valuable resource coveted for its prized bird shoots. Its protection and nurturance ultimately depends on the commitment of a series of individuals who love the lake and birds, from the game-watcher of Lungh to the senior-most provincial official willing to litigate for its tenure.

The natural history of Lungh locates it as an oxbow lake of the Indus River, an ephemeral phenomenon in river flood plains, that has been salvaged and managed as an important staging site for thousands of birds. These birds come to rest and feed in safety as they overwinter in Pakistan, where they arrive from Siberia on one of the great global migrations along the Indus Flyway. The open waters, swamps, and rice fields of the hundred-acre lake are all important elements of the panoply that is a paradise for birds.

Lungh is one of the forty plus lakes and marshes in Sindh alone that are a destination for birds making their annual passage on the Indus Flyway, one of the world's greatest bird migration route that passes through Pakistan. The story of Katerina, a black stork who was radio-collared by Czech scientists, tells of her arduous passage across the great physical barriers of the high Karakorams and down the Indus Valley, and the adventures that befell her as scientists tracked her resting places and journeys through Central Asia to northern Pakistan. Perhaps if Katerina had made it alive to the haven of Lungh, she would have been safe enough to survive the return journey to her home in the cold north of Siberia.

(facing page)
British India in 1870; officers of the Madras army seen during a duck shoot in the Nilgiris. Colonial imperialist character building was part of the Shikar culture associated with wetlands all over India. (By permission of The British Library, 447/23 Shikar Madras.)

One of the great global bird migration pathways, the Indus Flyway is the route taken by birds who overwinter in southern Pakistan before flying back to the northern Hemisphere for the summer.

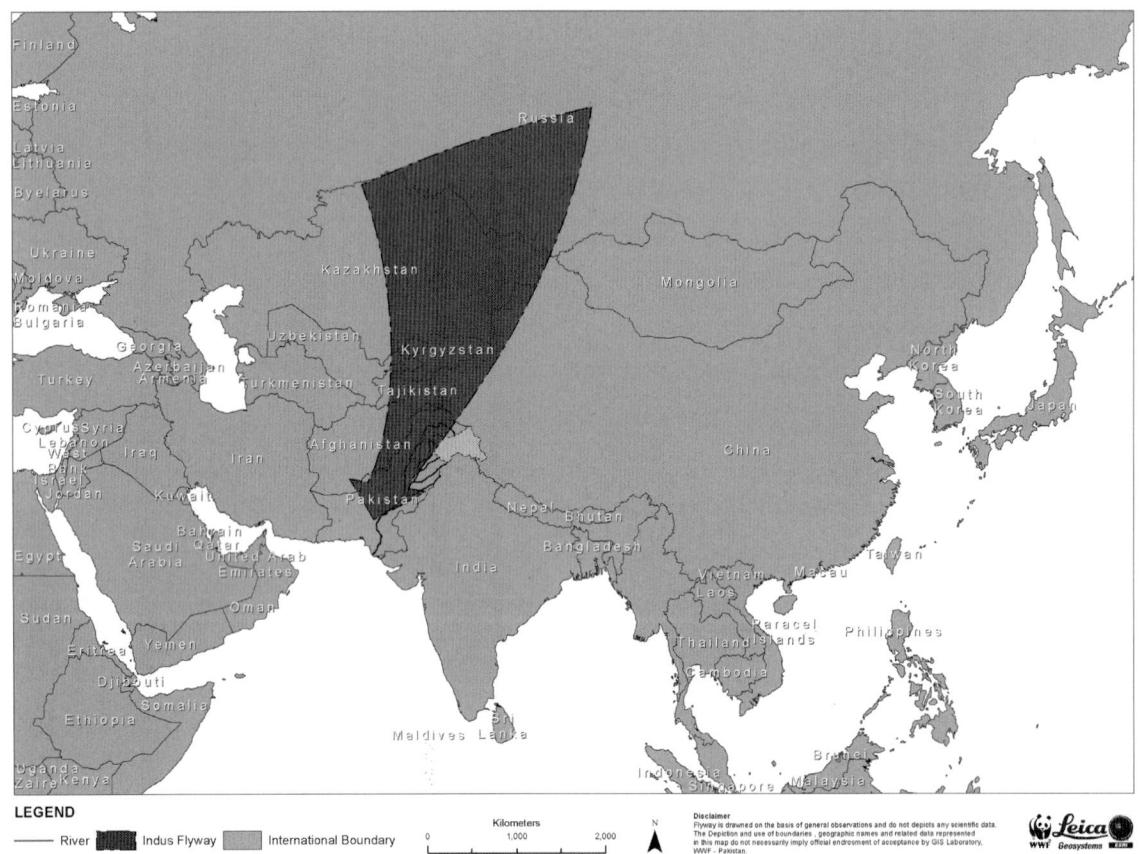

LEGEND

—— River ▓▓ Indus Flyway ▓▓ International Boundary

Kilometers
0   1,000   2,000

N

WWF *Leica* Geosystems

Migratory birds' food and feeding habits differ widely to avert competition. Each species has a special diet and way of feeding, reflected in the shape of their beaks.

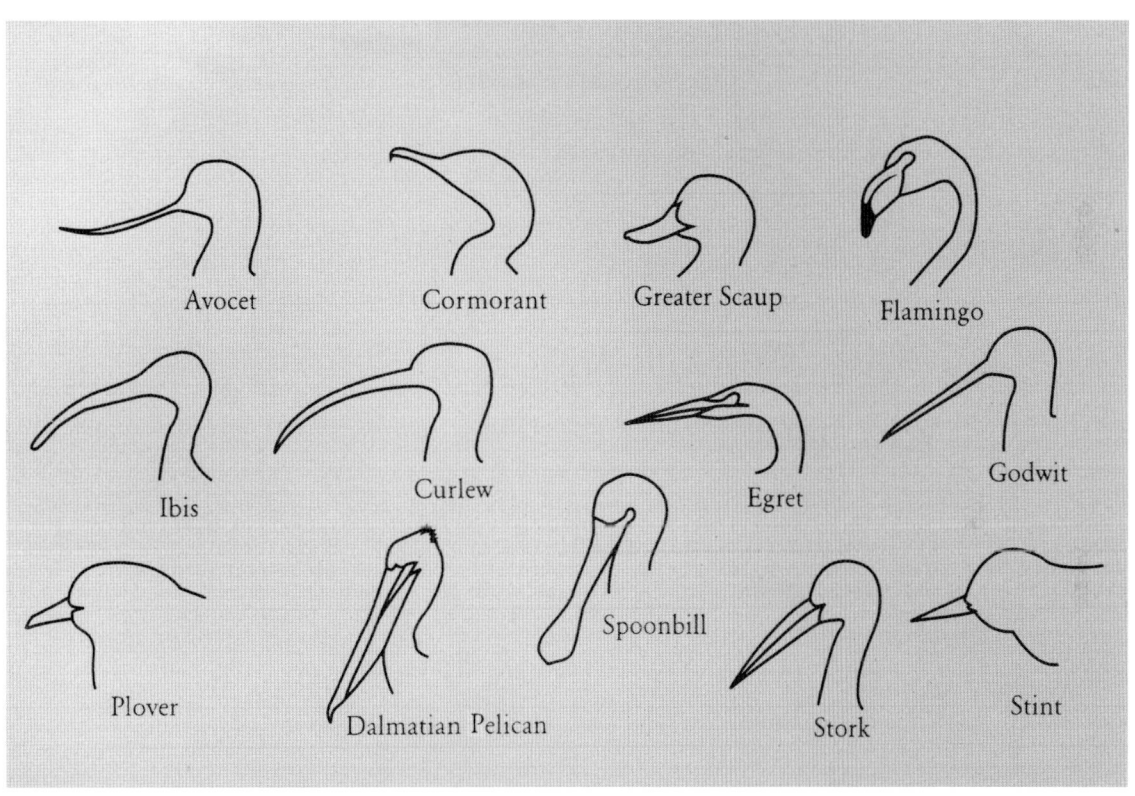

Courtesy: WWF-Pakistan

The striking Pintail drake is a dabbling duck that flies from faraway Siberia and is found as a winter visitor at Lungh Lake.

Courtesy: Ghulam Rasool

Lungh Lake — A Paradise for Birds

In the winter months, when thousands of migratory birds fly over the Indus, looking for comfortable places to rest in upper Sindh, one lake stands out for its sweet-waters, undisturbed typha plants, and swampy area overgrown with tamarix. This is Lungh Lake, a wildlife sanctuary located in the fertile flood plains of the Indus River in District Larkana in the biogeographical region of the Thar Desert.

The crescent-shaped oxbow lake was once a part of the Indus riverbed, which has now shifted 40 kilometres to the east of the lake. An oxbow is a small lake located in an abandoned loop of a river channel, which in this case was that of the Indus River. Lungh Lake is today an important wetland spread over nineteen hectares, hosting a large variety of migratory waterfowl on the Indus Flyway and is considered one of the best-maintained lakes in northern Sindh.

Lungh Lake is protected from the surrounding villages and agricultural fields by the Sindh Wildlife Department. It has been turned into a safe sanctuary for birds with a forty-seven hectare buffer zone. There has been no hunting on this lake since the 1990s, even though the nearby villages of Ghogari and Pakho fall within the radius of the sanctuary. This means about 12,000 people living in the surroundings of Lungh Lake are largely dependent on the lake's natural resources for grazing their livestock. Despite this, the local villagers do not hunt or disturb the birds resting on the lake. There are five game-watchers, two guards, and one assistant of the Wildlife Department for the protection of the wetland's resources.

Like so many oxbow lakes in the subcontinent, the land surrounding the lake and its buffer zone is flat and divided into small rice fields irrigated by a network of small canals. There is no outlet for the lake and the boundary is surrounded by a man-made *bund*, or causeway, which helps to retain the lake's fresh water. The average annual rainfall here is about seven inches, and most of it precipitates during the monsoon season in July and August. The climatic conditions of the area are extreme and the average temperature reaches up to 47 degrees centigrade in the summer months, falling to 5 degrees centigrade in the winters.

In the winter months, thousands of birds fly over the Indus and settle in Lungh Lake.

As a freshwater lake that is dependant on monsoon rains and surplus water from the surrounding rice fields, Lungh Lake dries up during the hot summer months. This is when irrigation water has to be supplied from the Rice Canal, which brings water from the Indus River near Sukkur Barrage. Over the years, however, the size of the lake has shrunk as less water is diverted from the canal.

Its importance as a bird habitat is in its open waters containing a dense growth of typha, and a swamp around the lake dominated by tamarix. The tamarix area is an important wintering ground for many smaller birds, including warblers. *Juncus* grass grows on the edges of the lake near the rice fields. Once part of the riverbed, the wetland contains fertile silt soils—comprising about 20 per cent sand and 80 per cent clay. The area of open water in the lake is around 20 to 30 metres across, and the lake's surface is dotted with birds in the winter. Each of these features in the natural layout of the lake are important to the micro habitats for the diverse migratory birds who find shelter and food here.

Lungh Lake is located at the confluence of the 'Green Route' or Indus Flyway. This is an internationally recognized migratory route followed by birds coming from the colder Central Asian regions to Sindh to spend winter every year. In recognition of the lake's importance as part of the Indus Flyway, in 1989, the Sindh

The elegant female Common Kestrel, a bird of prey that eats small rodents and lizards in Sindh.

Courtesy: Ghulam Rasool

The Curlew, with its long, curved bill, hunts on soft mud-eating worms and crabs.

Courtesy: Ghulam Rasool

Wildlife Department began to actively protect and improve the lake, turning it into an ideal refuge for migratory waterfowl.

The lake supports large numbers of waterfowl in addition to large populations of non-migratory resident species. The important birds found at Lungh include the Purple Moorhen, Great Cormorant, Common Teal, Grey Heron, Curlew, Lapwing, Avocet, Indian Shag, Spoonbill, Common Pochard, Coot, Marbled Teal, Mallard, Pintail, Tufted Duck, Moorhen, Gadwall, Gargany, Shoveller, Imperial Eagle, Common Kestrel, Common Shelduck, and Wigeon.

The sanctuary offers refuge every winter season to these birds, which come in autumn and leave in spring. December is the very best time to view the ducks, as their diversity, ease, and approachability offer a unique feast to the eyes. Here, researchers, bird-watchers, students, and wildlife lovers are encouraged to visit the lake and enjoy the wildlife without disturbing it. An educational and information centre facility is also available for visitors.

Amongst the larger mammals, the Asiatic Jackal, Red Fox, Wild Boar, and Jungle Cat have been spotted around the lake and its vicinity. A total of twenty-five fish species have also been recorded from Lungh Lake during recent studies and its fish fauna are similar to those found in the Indus River in the Sukkur area. This is because the primary water supply into Lungh Lake is through the Rice Canal originating from Sukkur. It is the waters of the Indus that carry fish spawn into Lungh, where they grow and multiply.

The lake is now over a century old, formed in 1880 from an abandoned river course of the River Indus. Oxbow lakes are generally formed as a river cuts through a meandering neck to shorten its course, causing the old channel to be rapidly blocked off. Isolated from the newly shortened river, the older route eventually forms an arc-shaped lake.

An oxbow lake may survive as a water body for some time, especially if ground water seeps into it. Alternately, the oxbow lake may dry up or fill with sediment, as was the case with Lungh Lake. In the past, the lake has faced both water shortages and siltation.

The construction of the Sukkur Barrage cut back seasonal water flows from the Indus River in the 1930s, leading to the dependence on water from the Rice Canal. In the 1980s, the lake became heavily silted and many birds stopped resting here.

Thanks to the efforts of the Sindh Wildlife Department, however, Lungh Lake has been regularly de-silted and maintained with enough water over the last two decades to retain its status as a birds' paradise. Today, it is an important site for waterfowl, offering staging and wintering grounds to over 50,000 water birds of numerous species. Accessible from Larkana town by a metalled road, it ranks amongst the top lakes for bird watching in Sindh.

The Kingfisher uses typha grass as a perch to watch the water's surface on Lungh,
waiting unseen to dive underwater for its prey.

A typical group photograph of a shooting party at Lungh circa 1970s. These shoots were symbols of government favour at a time when Pakistan was a young country.

Left to   Right:— Syed Daim Shah, Master Hisamuddin, Mr. Abdul Karim Shaikh, Mr. Roshan Ali, Shaikh, Sardar Mohammad Aslam,
S.P.   Gen: Secy, P.M.L.
Jam Sadiq Ali, Pir Shah Mardan Shah, F.M. Mohammad Ayub Khan   Mr. Mohammad Nawaz Shaikh, Kazi Fazlullah,
M.N.A.   (Pir Pagaro);   President of Pakistan   (Host);   Home Minister
Mohammad Aslam Bajwa, Mr. Mubashir M. Khan, Mr. Anis Arif, Mir Aijaz Ali, Syed Nadir Shah,
Commr. KHP.   D.C. LRK   DIGP   MNA   MNA
Sardar Ghulam Mohammad Mahar,   (Sind Photo Studio Larkana.)
MNA   Phone No 2229

Cultures that Shaped Lungh

Wildlife management institutions in the Indian subcontinent reach back to a time when they arranged Shikar for Mughal kings. They then evolved into colonial administrations with relevance to how they are shaped in Pakistan today. What are the lineages of these institutions, and how have they adapted from control and exploitation to conservation concerns over the centuries? Can we find examples of leadership that shaped colonial wildlife management regimes to protection and valuation through research, education, and recreation? The institutional history of Lungh is one such example, as it represents a loved and protected wetland site accorded effective public protection.

Mughal traditions of Shikar are evoked in the annals and miniature illustrations of a succession of kingly accounts of court life. In the 1500s, hunting was one of Mughal Emperor Akbar's well-known pleasures, reportedly keeping him fit and his senses alive. His Mongol ancestry brought with it a culture of kingly 'training for the eye and quickening of the blood'. Later, in the 1600s, Emperor Jehangir's hunting department grew into a fully fledged arm of his court; well-resourced, carefully counting and recording the hunts of the court to leave behind detailed accounts of large mammals caught as well as bird hunts for peacocks, egrets, and ducks. In this era, Shikar was also a woman's sport — Emperor Jehangir's wife Nurjehan was a notably sharp markswoman and huntress.

With the rise of the British colonial empire, Shikar continued to be a symbol of kingship, becoming a part of the construction of the British imperial and masculine identity; a prized 'character trait', so valued at the time. It was promoted as a deterrent from 'idle pursuits', and modern scholars also note its symbolic significance in the imperialist mind of dominating Indian nature. Through the 1897 Forest Act, the Indian Forest Department charged with wildlife management, became one of the powerful organs of the empire. Hunting licenses became social transactions and negotiations, jealously guarded for the privileged few. Records

On a cool winter morning, wetland reed and typha grasses are transported for use as screens in homes and for sale.

Migratory birds take flight on Lungh, a haven of rest and safety during the long winter migration from Siberia to the Indus wetlands.

show that Viscount Earl Mountbatten had a shoot at Drigh Lake, not far from Lungh, as did the Shah of Iran. These shoots were symbols of national favour at a time when Pakistan was a young country, vying to win the patronage of larger powers.

After the creation of Pakistan, the Sindh Wildlife and Forest Department, still managed under the jurisdiction of the Forest Act, established an innovative semi-autonomous body called the Sindh Wildlife Management Board. In the newly founded country's heyday, this body constituted a distinguished and hand-picked group of men chosen for their passion for nature, variously garnered through Shikar, love of field sports, science, and conservation.

Amongst other lakes in Sindh, they recognized Lungh as a place of importance both for bird shooting and as a nature reserve. It was accorded legal status under their leadership, and its management was supported to ensure that the lake remained a viable bird sanctuary.

The management of Lungh Lake as a sanctuary for migratory birds spans some 120 years, commencing in 1890 when it was recorded in colonial annals as a natural depression which was filled with waters east of the Indus River. By this time, it was probably still an oxbow lake that was cyclically inundated by the River Indus itself; but by the 1930s, when the Sukkur Barrage, one of the earliest great structures to intercept the Indus, was built, freshwater recharge to Lungh was stopped and water from the Rice Canal helped to recharge the lake.

It was in the 1960s that the lake was recognized as a hunting reserve where Shikar for birds was permitted for important personages such as Zulfikar Ali Bhutto, a native of Larkana District. To aid in this activity, some of the early bird counts of migratory fowl were carried out and still serve as useful records. Around a decade later, Lungh was formally notified in 1972 as a wildlife sanctuary under the tenure of W.A. Kirmani, a pioneering conservationist, initially for a period of five years only. In these years, the Sindh Wildlife Management Board (SWMB) laid down the core priorities and legal systems to protect habitats, and used its large jurisdiction to effect in Lungh. But as their strength grew, so did the political interest in the exclusion of lakes such as this from Shikar. The lake received a serious setback between 1977 and the 1980s as the powers of the SWMB fell into abeyance, and habitat degradation took place.

In 1982, it was finally notified as a wildlife sanctuary in perpetuity, with hunting and poaching of waterfowl strictly prohibited, and a ban on grazing, tree cutting, polluting, and fishing in the lake. By the end of this decade, the Sindh Government allocated development funds to the sanctuary, to de-silt and restore its habitat for birds with the construction of an embankment, bird watchers hides, and weeding the lake of excess vegetative growth. At this time, surrounding land to the order of some sixty acres was purchased from local farmers to make a buffer zone to extend beyond the core forty acres of Lungh.

Once again, the investment in and prohibitions imposed on Shikar invited conflict as the Sindh Wildlife and Forest Department faced litigation from the local Kitchi tribe, who claimed that the land and the lake once belonged to them and demanded compensation money. In an era when political forces favoured privatization of public resources, Lungh too was somehow leased out to a private party for management. This party was known to be under local political protection. In this instance, the government filed a petition in 1997 to have the lake restored to departmental management and won the case.

In the 2000s, Lungh remains a wetland sanctuary of significance that was periodically endowed with public money to maintain it as an important migratory staging site. Unlike others, this wetland is not immediately threatened with development works, but the threat posed by human beings is classified as 'lawlessness', which refers to the armed tribal feuds that are played out in the fields and roads in the environs of Lungh. As is natural to an oxbow lake, the overgrowth of grasses, the accumulation of silt, and the loss of freshwater inflow are a constant pull to diminish the lake. After all, the oxbow, left to nature, would eventually become a dry scar on the face of the land, a mere memory of when it once teemed with thousands of birds from the cold north, as Lungh continues to today.

Barkat Ali, seen at centre in white clothes, is the game-watcher at Lungh Lake.
He owns no high-tech equipment, but his intimate knowledge of visiting birds
indicates that 70,000 birds in December is a good year for migratory fowl.

Barkat Ali — The Game-Watcher

'There was a time in the 1970s when *Badshah* (King) Bhutto used to come here to hunt along with the likes of Sheikh Zayed and Yasser Arafat,' brags Barkat Ali, the all-purpose guide, wildlife guard, and cook at the Lungh Lake rest house. It is early afternoon and Barkat Ali is preparing lunch; his long shirt's sleeves rolled up as he blows into a small cooking fire. Barkat Ali, who comes from the nearby village of Ghogari, was himself born in 1982 but he seems to know all the grand tales about the old rest house, constructed in the 1970s, when the lake was used as a hunting reserve by the Government of Pakistan. The rest house is now owned by the Sindh Wildlife Department. 'Everyone's favourite hunting duck was the Mallard. They would kill dozens in one shoot alone. But since the 1990s, there has been no hunting on the lake. It also used to be bigger back then; now it has shrunk.'

Lungh Lake, its open waters covered by typha, is surrounded by rice fields with a few villages located just beyond. The lake has considerably shrunk in area as water from the nearby canal, which is used for irrigation, is not regularly diverted to the lake. In the thirsty province of Sindh, competition for water in the canals is a way of life. In Barkat Ali's view, a good year for the lake is when around 70,000 birds come to visit. Last year, he counted only 40,000 birds. He spends his day patrolling the lake, chasing off any trespassers, and occasionally feeding rice to the birds.

We give them feed to invite them to stop and rest here. Otherwise they will fly off to other lakes in the area where they can be hunted down. The weekly license for shooting at nearby lakes costs approximately two thousand rupees. This is a secure area for them and the local villagers don't disturb the birds. Most of the locals are aware of the lake's special status as a wildlife sanctuary. I, for one, live at the quarters near the rest house and am here twenty-four hours a day.

Barkat Ali has spent most of his youth as a game-watcher. As a young man, he received training from the Sindh Wildlife Department and says he has never hunted in his life. As he roasts the freshly-caught fish from the lake with coarsely chopped onions and tomatoes, he talks about the local Sufi Saint Syed Ali Shah, whose shrine is located nearby. 'Many women in the area visit the shrine to pray for children, and it is regarded as a special place where prayers are fulfilled.' The people of Sindh love their mystic saints who preached tolerance and respect for all of God's creations, and they regularly visit these shrines. Barkat Ali also goes there to pray for guidance from the saint, whom Sufis believe emits grace and intercession. 'I always had this deep desire to work for wildlife. I'm very happy with this job,' he explains. 'I am up at 8 a.m., spending the whole day walking around the lake, looking at the birds and making sure no one disturbs them. There are hardly any outsiders here and in all my years here I have never caught anyone trying to shoot the birds so no one has been fined. There are eight guards here in total, with two of us in charge. I know all the different ducks and bird species but I'm not well-educated. I only finished Class 5.' Barkat Ali points out a Moorhen, which nests here from August to September, and a Gargany Duck. Overhead, he spots an Imperial Eagle, while a Kingfisher darts into the *Juncus* on the edge of the lake. He says that the endangered Marbled Teal has not come in recent years to visit Lungh Lake.

Barkat Ali owns no high-tech equipment; he uses no binoculars or spotting scopes, but goes enthusiastically about his job, even spending the occasional night patrolling if need be. 'This is generally a very peaceful place. Guests come and go, and I show them around. Most visitors come in December when the lake is full of ducks, then I enjoy taking them around,' he says, as he stands in the midst of a throng of water birds, lake, and typha.

Lungh Lake in Larkana District is a paradise for migratory birds because a group of committed naturalists and scientists laid the foundation for bird conservation.

The protection accorded to Lungh Lake is a government success spurred on by a pioneering group in the Sindh Wildlife Management Board. The benefits of this haven for migratory birds are beyond measure.

The Indus Flyway

The ponds, lakes, and riverside pools that dot the Sindhi landscape are loved and nurtured by landowners and government alike, for they form a part of the cultural imagination and identity of the subcontinent at large, and this ancient province in particular. Anyone who experiences the unexpected encounters with migratory birds in the wilderness realizes it as a moment of unforgettable wonder.

The marvel of thousands of birds flying to southern Pakistan to herald winter has great cultural continuity, but is fraught with change as bird migrations become vulnerable to a variety of threats. The country is an active contracting member of two key international conventions that protect this natural phenomenon: the Ramsar Convention, which calls for conservation of wetlands, and the Bonn Convention for the protection of migratory birds and animals are both important instruments put into place by the international community.

There are seven identified flyways in the world representing north-south passageways where millions of birds migrate large distances across the globe. The Central Asian Flyway covers the areas used by species of birds from Siberia through Central Asia towards the Indian Ocean in the south. It overlaps with the African-Eurasian Flyways in the west and the East Asian Australasian Flyways in the east; with the meeting of these in two streams in Pakistan. This famous route from Siberia to destinations in Pakistan soars over the Karakoram, the Hindu Kush, and the Suleiman mountain ranges along the Indus River down to the delta and coast. This is known as the International Migratory Bird Route Number 4, or the 'Green Route', and more commonly, the Indus Flyway.

It is estimated that over a million birds from the Siberian region fly to Pakistan every November and 70 per cent of them overwinter in the safe confines of wetlands that are declared Ramsar sites in the country. Endangered species of repute include the White-headed Duck, the Houbara Bustard, and the Siberian Crane. Sindh alone has some forty-five such sanctuaries, altogether covering just under ten thousand square kilometres. Haleji Lake, near Karachi, teems with migratory birds. When the Duke of Edinburgh Prince Philip visited the lake in the 1980s, he pronounced it a bird watcher's paradise.

The challenges of securing nature's passageways defy human-made political boundaries so that protecting global bird flyways has come under recent scrutiny as an opportunity for international cooperation. Threats to flyways include the expansion of agricultural lands and hunting, including sport hunting which, once a cultural activity for few, has now reached dimensions that threaten the populations of many migratory bird species. Larger birds, such as the crane, are also liable to collide with human constructions such as transmission lines. But the opportunities to secure these flyways are many, especially in Pakistan, where the pride and love of waterfowl is common. The protection accorded to Lungh Lake is a government success spurred on by a pioneering group of nature lovers in the Sindh Wildlife Management Board. Its repercussions for birds are beyond measure.

We believe the sad tale of Katerina* the Siberian Black Stork, would have had a very different ending if she had reached the sanctuary of Lungh. She would have been among a dense family of diverse birds who rest here in safety, sustaining themselves before their epic journey back to their summer grounds in the North. We know that individuals like Barkat Ali, the game-watcher of Lungh, would have cast his protective eye on the little lake and the stork, and played his part in the marvel that is the passage of a bird migration, true to the symbolic connections that his Sindhi heritage instructs as the meaning of the soul's journey through this world.

---

*Katerina was a tracked migratory bird, hunted as prey by locals. See page 151–6 'Story of Katerina', for further details.

The Grey Heron, a native breeder in Sindh, is an elegant bird which takes flight slowly and retracts its neck into an artistic S-shape as it becomes airborne. Often found in oxbow lakes such as Lungh, it feeds on mollusks, insects, rodents, and crabs found in abundance in its shallow waters.

Photograph courtesy: Ghulam Rasool, by permission of WWF-Pakistan/PWP

*In the Siberian forests is the summer home of the Black Stork named Katerina. Near the
Ob River of Novosibirsk is her nest, which she will leave as winter approaches to fly
the incredible journey to Pakistan and overwinter in the Rann of Kutch.*

All photographs of Katerina by African Odyssey Project

From Siberia: The Story of Katerina*

very year, billions of migratory birds leave their nests in the Palaearctic region and head south to their wintering grounds. Which way and where exactly do they fly? How do they cope with different geographic obstacles? Where and for how long do they stop? How do they navigate? What endangers them along their journey? How do we protect them more effectively? After nearly a hundred years of research, we are still unable to find complete answers to these questions. Today's advanced technology, however, enables us to get new findings quicker and get closer to answering at least some of the questions. At the same time, it gives rise to new questions. We can illustrate this with the story of Katerina, the Black Stork.

Satellite telemetry changed the approach to researching bird migration. Birds carry miniature transmitters, whose signals are detected by satellites running on earth's orbit. The researcher receives locations of the tagged individual directly into his computer. This development of satellite telemetry inspired many research teams into monitoring migration of different species of birds. In the mid-1990s, our Czech team was inspired by the new opportunities available and decided to tackle the financial limitations with enthusiasm. The African Odyssey Project was launched. During the course of the project, almost twenty Black Storks (*Ciconia nigra*) were observed migrating from their central European nests into their wintering grounds situated in tropical Africa, in the region between the southern edge of the Sahara Desert and the Equator. Within the next five years, some of the storks equipped with transmitters were repeatedly monitored, and expeditions were organized to their stopover sites and wintering grounds. A female named, Kristyna, was followed by the 'step-by-step' method all the way from the Czech Republic to Senegal. The African Odyssey, closely followed by the public, brought many findings, some of which widened the mosaic of our knowledge, while others had a practical impact on the conservation of migratory birds.

---

*By Miroslav Bobek, New Odyssey Project.

Although the African Odyssey ended in the year 2000, its legacy continued, resonated by the words of a Czech proverb: 'the more you eat, the more hungry you get'. By the year 2002, a Czech team was operating in Siberia on the project 'New Odyssey'. This had similar aims as its 'African' predecessor, but was situated in the vast continent of Asia. Gradually, the Black Storks would be monitored in different parts of their nesting habitat — from western Siberia to the Far East. It was a mission for many years to come, as well as a difficult task complicated by many unfavourable and unexpected circumstances. That was proved at the beginning of the project in summer 2002. Information about stork nests obtained during spring from our Russian colleagues was incorrect and we had no other option than to search in the taiga around Yenisei and in other parts of Siberia. For two weeks, the team worked in a large area determined by the cities of Krasnoyarsk, Kyzyl, and Novosibirsk until it managed to find the first occupied Black Stork nest. It was found near the Ob River in Suzun District, south of Novosibirsk.

Subsequently, the team was able to catch and equip with a transmitter its first 'Asian' Black Stork. Two more nests and two more storks had to be found. Time was running out fast and the date of departure back to Europe was approaching. After finding the second nest and marking the second individual with the help of local people, the team managed to find (only 36 hours before departure from the region) the third nest. A female was tagged there and she was named after the legendary Russian ruler Katerina (Catherine the Great). From there unfolded a story which seemed to end half a year later in northern Pakistan. In fact, it continues until today.

So far, the Czech team has followed the migration of five storks coming from the Ob River in southern Novosibirsk region. The results of this observation suggest that migrating birds usually set off in the south-western direction, thus avoiding the Tian Shan, Pamir, and other mountain ranges. These storks then halt for a few weeks near Amu Darya or Syr Darya Rivers and then fly to the south-east and through the Afghan Hindu Kush mountain range to the Indian subcontinent. Only a small number of Black Storks fly along a different, much more difficult route — a route which came to be known as 'Katerina's route'.

Katerina left her nest in August and headed straight to the south, directly towards the world's highest mountain ranges, and stopped only after reaching the Dzhungar Basin in north-western China. From there, she flew in a west-south-westward direction. On 9 September 2002, she found herself in a deadlock between high peaks near the Issyk-Kul Lake in Kyrgyzstan. She backtracked in the following days along an almost identical route and then rested for about three weeks in south-eastern Kazakhstan. She then set off again for Issyk-Kul Lake. This time, she decided to fly on and in two days, between 30 September and 1 October 2002, she crossed the Tian Shan. She changed her direction towards the east and reached the Tarim Basin in western China. She stayed there for a long period — more than ten weeks. In the last 3–4 weeks of this period, she kept flying from place to place, probably searching for a climatically suitable and food-rich location. Everyone who followed Katerina's journey on the Internet started to get worried. She came close to the Kunlun Shan at the southern edge of the Tarim Basin twice. It seemed that she wanted to go further south to the Indian subcontinent. Anyhow, she was separated from it not only by the Kunlun Shan but by the Karakoram Range as well. It seemed almost impossible that she would be able to cross a mountain range whose integral part is the second highest mountain in the world, the K-2, especially with winter arriving. But this is exactly what happened. On the third attempt, she crossed the Kunlun Shan on 14 December 2002, and on this very same day reached the Khunjerab Pass (4,730 metres) in the Karakoram Range. The next message from her transmitter revealed her location in northern Pakistan. She was literally in the shadow of Nanga Parbat. She crossed the cold mountains, where she couldn't be supported by rising air currents but instead had to face harsh gusts of wind. Her performance couldn't be considered anything other than fascinating. Was this anything but an accidental aversion? Was it possible that the same route was used by other storks?

The answer to all these questions arrived sooner than any of us expected. Satellite data from Katerina's transmitter around Christmas started to worry us. It showed that the transmitter was stagnant by the Indus River near a small town called Chilas. It was not changing its position—the information from the movement sensor was constantly on the same level. Katerina's temperature sensor was showing abnormal temperature averages; in the evenings, the temperature would uncharacteristically go up, reaching up to 20 degrees centigrade, instead of declining. As if the transmitter was in some house, where people would start cooking in the evenings...

One month later, two jeeps with panda logos on their doors were approaching Chilas on the Karakoram Highway. Pakistani nature conservationists reacted extremely quickly to the news of presumed death of Katerina and offered cooperation to the New Odyssey team. In the jeeps approaching Chilas—and Katerina's transmitter—were Czechs along with employees of WWF-Pakistan. With every kilometre, they grew more and more tense. Will they be able to retrieve the transmitter and find out the fate of Katerina? There could be a problem with the local people—proud, unapproachable, and armed. Initially, the cars drove across the location where the transmitter was last recorded by the satellite without even slowing down, but then the antennae hidden in one of the jeeps caught the signal of the transmitter. It was obvious that the transmitter had to be in one of the houses in a settlement a few kilometres behind Chilas. Now everything was in the hands of the employees of WWF-Pakistan. Thanks to their skilful negotiations, our international team could enter the settlement and start a long discussion with its inhabitants. At the end of the discussion, it was clear as to what had happened to Katerina:

It was late in the evening and slightly drizzling. Little Samiullah was roaming around the banks of the Indus River when approximately five hundred metres upstream on a sand bank he saw a number of large birds. He immediately ran home. The first person he met was his older cousin Ziauddin. He took an old German shotgun and went along the road to the sand bank. When they were almost above it, they cautiously came down the cliff hidden by large boulders crawling towards the water. Finally, they were close enough. A shot was fired. About five or six storks flew off while one stork stayed lying on the river bank. Ziauddin came closer, picked it up and only then he saw that there was something fixed on the stork's back. A transmitter...

There is no reason not to believe his account that there were storks in addition to Katerina. This is extremely interesting as well as important. Katerina did not fly across the Karakorams alone and her route — as complicated as it seems — was not a route taken by chance. The work of the Czech-Pakistani team increased the knowledge about the fact that there is an important migratory route through the Karakorams and the Indus Valley. Apart from being traversed by Black Storks, it is important for other migratory birds as well. They are all intensively hunted with the exception of birds of prey. From there, it was only a small step away from the idea of trying to do a project which would bring the research on migratory birds in this region together with an attempt to protect them. This would include spreading awareness among the locals. Realization of this project would be the best end to Katerina's story.

Even though observing Katerina answered some of the questions, further questions arose that needed answers. The surprising development of her migration became a strong incentive to refine thoughts on navigation mechanisms in Black Storks and the method and extent of their migratory routes. The members of the New Odyssey team hope that they will tag a 'new Katerina', a Black Stork which will fly through the world's highest mountain range; a natural laboratory for researching different aspects of bird migration.

There are other important questions that remain. Where and which way would Katerina have flown if she wouldn't have been spotted on that rainy day in December by little Samiullah? Would she have gone to south Pakistan to the lower reaches of the Indus River, or to the marshlands near Karachi, or would it be India? Near which water sources would she have stopped? Would she be safe on her stopover sites and in her wintering grounds? At least some of these questions can be answered by further satellite monitoring of Black Storks and by the patient fieldwork of ornithologists and nature conservationists equipped only with binoculars.

Czech scientists locate Katerina in the Siberian wilderness and fit her with a satellite transmitter, an instrument that allows tracing her migratory path to Pakistan in great detail.

*Katerina made the incredible migration across the Kunlun Shan into the Karakorams
to reach Khunjerab Pass in Pakistan. She settled in the Indus Valley at Chilas,
where an expedition of scientists located her satellite transmitter in the
house of little Samiullah, close to where she had been shot.*

*The Black Stork's winter home in India, located in the mild coastal inlands of the Rann of Kutch. This would have
been Katerina's destination had she made it alive across Chilas, then rested in the sanctuary of Indus lakes
such as Lungh, and arrived in the fecund marshes of Gujarat.*

Introduction

Arising from the marvellous irrigation works of the early decades of Pakistan's independence, parts of the Thal Desert were provided with water through the Muzaffargarh canal. In the sandy soils of this central desert of Pakistan, a place in between the great rivers of the Indus Valley, water seeps through to form ponds, marshes, and moorlands of sedges in natural depressions. Rangla is one such place in central Punjab, a wetland formed unplanned that has become a place where nature and people have settled into a state of harmonious coexistence, spread over an area of eighteen marshlands.

Rangla caught the scientific imagination when the endangered Marbled Teal, a shy and secretive migratory bird, was first seen there some two decades ago, reportedly with a little clutch of nestlings. This observation alerted conservationists to the reality that Rangla had become a sanctuary and breeding ground for important avifauna and needed protection from hunting and habitat destruction by unregulated visitors and local residents alike. A destiny-changing notification of Rangla as a Ramsar site of international concern may yet happen after many years of efforts by scientists, NGOs, and government to protect it.

An essay details the history of how the Thal was formed by river depositions over the centuries; how wetlands such as Rangla develop, setting both an ecological and social context for wetland and reed-dependant assemblages of humans, plants, and animals. The double threats of salinity and waterlogging that arise from the nature of the geographical depression and sandy soils brings to the fore the question of the future of such marsh-moorlands. Yet, as incidental as Rangla's history may seem, its people and plants are set in the historical context of this desert.

A focus on the Peelu bush of Thal Desert shows the iconic meaning of this plant and its berries, both instrumental and symbolic. Remains of the Peelu have been found in the archaeological digs of the nearby Harappan Civilization of 2000 BCE, underlining

(facing page)
*In southern Punjab, the culture is a bond between people and the austere desert environment.**

*All photographs in this chapter by Javaid A. Khan unless otherwise indicated.

The Thal Desert was formed by river depositions over the centuries and is home to clans of people who are traditionally camel keepers and woodcutters.

its ancient usage as a source of food for animals and humans across millennia. We find references to Peelu in the Hindu epic *Mahabharata* and later in the Holy Quran, and we deliberate on the *kaafi*, a classical form of Sufi poetry, of the nineteenth century patron saint of this desert, Khwaja Ghulam Farid, who used it as a metaphor for the seeking soul that resonates deep in the heart of the desert dweller to this day.

The people associated with Rangla are sharply separated by caste and wealth. On the one hand, the powerful and now political Khar family who own large lands in the region, and on the other, the camel keepers and woodcutter clans who now eke out a living selling reed products from the marshes. Although appeals have been made to members of the local elite to protect Rangla's wildlife, soliciting hunting on the wetland has profited the local communities, as affluent city dwellers come for seasonal duck shoots.

Our final cameo focuses on Tanvira, the huntsman of Rangla, and describes his meagre means as he illicitly shoots a duck to bring to his family's evening meal. Later in the winter months, he assists a local doctor from Multan to shoot ducks at Rangla in exchange for a reference chit that will allow prompt attention for his family at the district hospital.

Reed products form the essential materials of life near Rangla, including house thatching, string beds, and animal trappings.

Marbled Teal in flight at Rangla. The sighting of these birds changed the conservation destiny of these marshlands.

Courtesy: Ghulam Rasool

A Conservation History Matures

'The use of nature is a requirement of life, but overuse is disrespectful of nature', explains erstwhile Inspector General of Forests of Pakistan Abeedullah Jan. This is the ethic that has brought the conservation interest in Rangla marsh and moorlands, or the Rangla Wetlands Complex, as scientists know it, repeatedly to the fore since the early 1990s. As a wildlife reserve, it remains a partially protected area that returns again and again to the attention of bird and conservation specialists within and outside Pakistan.

Even though the Rangla marshlands are an incidental, man-made phenomenon, a series of incremental investigations have led to a deeper understanding of its conservation significance and potential as an ecology that merits protection. If it qualifies to become a designated Ramsar site, it has a special chance at attaining both ecological protection measures as well as enhanced development efforts. The Ramsar convention has developed a concept of 'wise use' closely echoing the sentiment of the Pakistani scientist quoted above, as it entails efforts to maintain ecological characteristics of a particular site, within the context of sustainably servicing the people whose lives depend on its resources.

Formed in the 1960s as an outcome of seepage from an irrigation canal built to bring water to the Thal Desert in the Muzaffargarh District of Punjab, Rangla caught the scientific imagination when a young zoologist sighted a very rare duck there. A hunter and bird enthusiast with medium-sized landholdings in the vicinity, this young man was a resident of Muzaffargarh town and worked in WWF-Pakistan. Together with his academic colleague in Multan, he sighted the Marbled Teal with a suspected clutch of nestlings at Rangla in 1993, and reported the sighting in the international newsletter of the World Wetlands Trust, along with the presence of several other endangered and endemic species such as the Sindh Jungle Sparrow and the Rufous Vented Prinia. A sighting such as this is capable of changing the conservation destiny of a location as it indicates a species of special interest, and one that is often considered a sign of the health of the area ecosystem.

Dr Aleem Khan, the young zoologist who sighted the very rare Marbled Teal with a clutch of nestlings — a sighting that caught scientific imagination and brought Rangla fame.

Following this discovery, the Ornithological Society of Pakistan, an NGO, was established with a particular interest in developing a management framework for Rangla, and also becoming the local representative of Birdlife International, an NGO based in the UK. The United Nations Development Programme (UNDP) and the Punjab Wildlife Department assisted the preparation of the first detailed plan for the Rangla marshes as a team effort constituting a group of men with a close affinity arising from student–teacher relations of the past. However, as persuasive and detailed as this plan was, its proponents scattered, as one team member retired from civil service and another migrated for graduate work, leading to an eventual emigration away from Pakistan. Despite the dispersion of those who conceptualized it, the community-based management plan for Rangla Wetland Complex of 1999 remains a seminal document that has reminded conservationists in Pakistan that the importance of this site cannot be forgotten. It sets the stage for asking a few essential questions for the 'wise use' of fragile areas, recognizing that whilst a traditional tenural regime of control by the local landlord or Malik may defy modern ideas of social justice, it is a reality that has served the ecology well. In a process of controlled hunting and lake management, development projects have been invested in the locality, as local dignitaries have been allowed to hunt here in exchange for the allocation of these projects. The management plan is based on the proposition that 'wise use' is compatible with traditional concepts of natural resource management, and may indeed benefit from being combined with a science-based approach to management. For its time, this was an idea well ahead of mainstream conservation frameworks in the country, and called for a Rangla National Park, a category reserved for areas of the highest ecological value.

Half a dozen years later, around 2005, the next major impetus in the conservation history of Rangla was the emergence of the Pakistan Wetlands Programme, a collaboration of the Ministry of the Environment and WWF-Pakistan. This long-term project carried out extensive baseline surveys covering many important wetland habitats throughout the country including Rangla. The overall findings of the surveys reveal sixteen threatened or near-threatened bird species in the country, recorded at fifty different sites. It identifies national priority action that should be given to some of these, including the Marbled Teal and the Ferruginous Duck, and refers to the latest April 2012 sighting of nineteen adults of Marbled Teal at Rangla, and later in June, of thirty-two individuals and a clutch of nestlings too.

As the two-decade history of Rangla marshes matures, Pakistan, as a signatory of the Ramsar Convention, has submitted a country report to its Conference of Parties in 2012. Five wetlands are in preparation for nomination as Ramsar sites, priority places of national action and international cooperation, and Rangla has finally made it to this list.

Young men hunt duck at Rangla; the wise use of the wetland means that controlled hunting may be possible on the lakes.

The Marbled Teal at Rangla, which is thought to be its sole breeding site in Pakistan. The inaccessible reed beds of the lake offer it courting and nesting grounds.

Courtesy: Ghulam Rasool

The Shy and Secretive Marbled Teal

Amidst the Thal Desert's wind-blown sand-dunes lies the inaccessible reed covered lake of Rangla, where the Marbled Teal can safely lay its eggs. The fragile ecosystem of the surrounding Thal Desert is so far intact and unspoiled by unwise developments and the shallow, brackish waters of the lake are rich in vegetation, providing the perfect habitat for the Marbled Teal.

Rangla Lake is said to be the sole breeding site of the Marbled Teal, which has been classified as a vulnerable species in the Punjab. The Marbled Teal is a shy, almost silent duck that usually lives in pairs or small parties. This silvery grey duck avoids open water, keeping to reedbeds and to stretches of water overgrown with weeds and plants where it can easily find concealment. It can fly well, but it is not as fast as the more powerful ducks. It can swim and dive with ease and feeds by dabbling, mainly in the top 20 centimetres of the water layer and, in autumn and winter, chiefly at night.

The Marbled Teal's numbers have declined in recent years due to unscrupulous killing by hunters from large towns who descend on the Rangla wetlands, comprising over a dozen lakes, for winter shoots. In the Punjab province, the Marbled Teal is mainly found on its autumn passage, although it is a resident bird in many lakes of Sindh. The vernacular name for the Marbled Teal in Sindh is 'Choi' and it is also found in countries north and south of the Mediterranean Sea: in the east, it is found throughout south Russia, Turkey, Palestine, Iraq, Iran, Afghanistan, India, and Pakistan.

Marbled Teals breed during the early summer and in Pakistan courting and pair-formation have been observed between April and May, while nesting itself is apparently variable in timing, generally falling between late April and the first half of July. Their breeding in various lakes depends on the rainfall and it is possible that in years when rain is scanty they move to places where water is plentiful. Their nests are made of rushes and weeds, usually concealed inside a clump of vegetation with an entrance tunnel roofed over by grasses in swampy land. Some of the places in

The Ferruginous Duck is another rare migratory bird found in Rangla. Few hunters are aware of its threatened status, making them a major threat to its population.

Courtesy: Ghulam Rasool

which they breed regularly are so remote that they are seldom visited.

The Marbled Teal's breeding in the remote wetlands of Rangla was first discovered in March 1993 when a hunter reported that he had seen a Marbled Teal with ducklings on one of the lakes. A survey was conducted by WWF-Pakistan in May 1993 and twenty-seven Marbled Teal were seen, some scattered in pairs while a few were found in small flocks. Although nests were not spotted, these observations strongly suggested that up to thirteen pairs of Marbled Teal were breeding in Rangla. It was the first ever record of Marbled Teal breeding in Punjab.

In a more recent survey conducted by WWF-Pakistan in June 2010, a flock of thirty-two individuals including six nestlings of about one week old and two empty nests were observed in Rangla. The Rangla Wetlands Complex is also an important breeding site for other waterfowl species, particularly the rare Ferruginous Duck. The main threat to waterfowl in Pakistan is from hunting and the Marbled Teal is considered a relatively tame duck that is easy to shoot, particularly in the breeding season. Few hunters are even aware of its threatened status.

The Rufous-Vented Prinia is another endangered bird found at Rangla, living in its reed beds.

Courtesy: Ghulam Rasool

Thal is an ancient desert of mobile and stable sand-dunes lying between two rivers, the Indus and the Chenab.

Water Seeps into an Ancient Desert

Dotted in the central deserts of Thal, Cholistan, and Thar are incidental by-products of the extensive irrigation works laid down in Pakistan over the last sixty years. These are marshes and lakes created by the seepage waters of canal systems and placed upon the sandy deserts created by the shifting courses of the mighty rivers of the subcontinent. Here, people and natural assemblages have woven their appearances in recent and elusive settlements.

Lying between two rivers, the Indus and the Chenab, Rangla is located in the Thal Desert where there is a natural depression near Panjnad (which means 'confluence of five rivers') as the rivers meet on their journey to the sea. This desert is one of the four major desert systems of Pakistan, which include the Thal, Cholistan, and Thar, often considered arms of the same alluvial sedimentation zone deposited over the millennia by winds and water from the north. The western desert of Kharan is unconnected to the above three deserts and offers its own unique continuities.

In these central Indus regions, irrigation for food production has been practiced for over three millennia as part of the earlier Harappan Civilization. But after the formation of Pakistan, the country had to harness enough water from the Indus River system to feed an entirely new nation, and look into the needs of the burgeoning population of the future. In doing so, Pakistan laid down a modern irrigation system that became a marvel of the engineering world, defying technical estimations of the scale and challenge of the works to become an international showcase of the 1960s.

As part of these marvellous irrigation works in the 1960s, the Muzaffargarh irrigation canal diverted water from the Indus at Taunsa Barrage and brought water into previously uncultivated Thal Desert region. Over the course of time, water from the two ancient rivers, the Indus and Chenab, had already seeped into the sandy soils of the area and created a high subsoil water table. When the newly conceived canal system was built, water began to seep out of the canal and form shallow lakes in the natural depression where the wetland complex of Rangla is located today.

Buoyed up by the subsoil water, the lakes have become a wilderness area in the midst of desert. Here, undisturbed by roads and construction, birds flock to the lakes and surrounding bushes at dawn and sunset. The Rangla Complex of marshes and moorlands located in the Thal Desert has been designated a wildlife reserve as it represents a special type of wetland habitat, distinct from the open surface lakes to be found in other areas of Punjab.

Situated sixteen kilometres from Muzaffargarh town, the total area of the Rangla Complex is between a hundred and a hundred and fifty square kilometres. Within this is an assemblage of eighteen marshes which emerged in the 1960s, and were scientifically discovered in the 1990s. The landscape setting is mobile and stable sand-dunes which make for a spectacular desert habitat of natural landforms. These are marshlands and overgrown lakes that provide refuge to birds where they can build nests, sheltered from predators by grassy roots and clumps on the water's surface. On the fringes of the lakes are assemblages of salt-tolerant native plants and trees; beyond are slacks or stable dips of sand in between sand-dunes that trap moisture and stabilize desert shrubs that offer a haven to various species of small mammals, such as jungle cats, porcupines, and wolves. The dunes themselves also shelter native clumps of grasses that invite many reptilian species.

Water in the Rangla marshes either seeps into them from the canals or enters as surface run-off received in the monsoon when the tributaries of the Indus are inundated. When the earthen banks of the nearby canals are broken, this also allows water to enter. As part of community efforts to manage the marshes, a nearby canal, One-R, or 'Wunhar' as it is locally known, has been broken deliberately to allow freshwater inflow. Accompanying this water, the Rangla marshes receive a rich supply of fish, fry, and fingerlings or fish eggs, from the Indus to populate its water.

The marshes near these freshwater inlets in the western edge of the cluster are mostly sweet-water lakes, and those on the further edge east into the Thal Desert contain increasingly salty water. This

*As part of community efforts to manage the marshes, a nearby canal, One-R,
has been deliberately broken to allow freshwater inflow into the lakes.*

The Muzaffargarh canal system allows freshwater to be diverted for release into the Rangla marshes.

spectrum of sweet to salty water creates a diversity of plant, bird, and fish life which offer considerable diversity within the wetland complex. For example, among the eighteen, the lake called Rangla is the most saline and hosts the enormous Tilapia fish, which grow exceptionally large as they live without competition for food with other species, being the only ones to tolerate salt levels this high. In turn, this saline environment invites cormorants, pelicans, and egrets, all of whom prefer this brackish water as they fly through on their spring and autumn migrations of the Indus Flyway.

Recent times and human activities have, incidentally, resulted in the formation of the Rangla wetlands; but these are also causing changes in the natural landscape and the marshes. The canals and marshes are changing the dune ecology by stabilizing some dune zones but enabling other zones to be belts of moving dunes to the east. Scientists do not yet understand these changes and the long term effects on this desert landscape are uncertain. The effects of human settlements near the marshes are also creating change: due to high ambient temperatures, the waters of the lakes are already low in life-giving dissolved oxygen levels but the process of eutrophication, or the decomposition of nitrogenous materials in the lake bottoms from human waste dumping, may also be causing the diminishment of oxygen levels. Nearby, the cotton growing regions bring chemicals into the waterways, and the extensive harvesting of typha grasses for use as fuel in the brick kiln industry can threaten this delicate ecology. Reducing the reed cover of the marshes can spoil the natural refuge that these lakes present for shy birds like the Marbled Teal, and also lower the buffering capacity of the lakes to periods of drought or flood.

Yet in this newly created marshland, there is much continuity of the desert culture and the connections that the people of southern Punjab have always recognized as a bond between people and the austere desert environment. The story of the Peelu tree shows it as an icon of this landscape and traces a historical, cultural, and poetic significance that is unfading.

The Marsh-Moorlands of Rangla with vegetation that offers a haven to its birdlife.

Photograph courtesy: Ghulam Rasool, by permission of WWF-Pakistan/PWP

*In the stillness of Rangla's wetlands, a Stilt wades to search for its meal in the shallows
with its highly adapted long legs, wading feet, and long beak.*

Photograph courtesy: Ghulam Rasool, by permission of WWF-Pakistan/PWP

A local Sufi shrine in the Rohi, as the Thal desert is affectionately known to its people.

The Peelu Berries Ripen ...

'*eloved show Yourself to me today, for the Peelu berries are ripening*' sings the voice of Reshma, a singer known as Pakistan's nightingale of the desert, herself once a desert dweller. This is a *Kaafi* composed by the patron Sufi saint of the central deserts, Khwaja Ghulam Fareed (1841–1901). A Seraiki speaker of Cholistan, or Rohi as the desert is called, he was a master of its geography and history, weaving his mystic metaphors with images of the desert that stir the hearts of his listeners to this day. In this instance, the Peelu berry is used as a metaphor for the ripening heart of the Sufi, which prepares to meet its Creator or Beloved, longing for gnosis in the midst of the worldly, desert emptiness outside him.

The Peelu, *Salvadora oleiodes*, is a small tree distributed widely in the desert regions of the Indian subcontinent and southern Iran known locally as *vann, jand, jar* (Sindhi), or *jall* (Seraiki). With yellow berry fruit that ripen in May and June, the Peelu berries can be dried and preserved in large quantities to be used as stores of nutrition in times of drought. Though slow-growing, it thrives in arid areas withstanding great soil salinity, withering when humidity in both soil and air rise above a certain level. Its dense shade and drooping canopy provide that rare facility in desert life for humans and animals alike: shade as well as excellent fodder for camels and goats. But it is its root systems that are noteworthy, as scientists have realized that its woody roots bind desert sand and enable other plants to grow in its shade; it is indeed a benign presence in the desert harshness.

Anthropologists believe that the relationship of this native plant and humans may date back to the Harappan Civilization of the Indus Valley region, whose ruins are found in nearby parts of Thal. Remains of this five-thousand-year-old metropolis reveal signs of Peelu collection in prehistoric households.

The Hindu epic *Mahabharata*, which was compiled two thousand years ago, makes mention of the Peelu, 'in forests having many pleasant paths of Sami, Pilu, and Karira'.

The Holy Quran refers to a related species, known to science as *Salvadora Persica*, more commonly found in Arabia and called *miswak*, a tree whose branches are used as toothbrush. A Hadith cites the Prophet of Islam: 'if it was not a burden for my adherents, I should make it compulsory to practice *miswak* after every *Salat* (prayer)'. (Abu Hurairah)

The flowers and berries of Peelu provide herbal medicine for *mange* (a skin disease) in camels, and the roots of the Peelu are considered the best toothbrush of all. Owing to these beneficial properties, the commercial trade of its root is causing the overuse of Peelu and may endanger the survival of the plant in the Rangla region. Mounds of unearthed sand are often to be seen around the tree base where woody roots have been harvested for sale as *miswak*. Science has found that the *Salvadora* genus does contain antibacterial chemicals as well as *salvadorine*, which form a collection of compounds that are beneficial for human gums and teeth.

*The Peelu berry is native to this desert and is used as a metaphor
for the ripening heart of the Sufi. Scientists also recognized it
as a benign botanical presence in the desert harshness.*

W.H. Fitch. del. et lith.

M. & N. Hanbart. imp.

Salvadora oleiodes, Decaisne.

The local communities living in Rangla are mostly from the Khandoya tribe,
who were historically woodcutters by profession.

Depending on Rangla for Prestige and Provision

Since the Rangla Wetlands were formed in the 1960s, they have been managed by the local chief or malik of this remote area of Punjab. The feudal system is still in force here and in recent years, the help of Malik Sultan Khar, who hails from the influential Khar tribe, was enlisted by the government to ensure that the Rangla Wetland Complex would be protected from unauthorized hunting.

Sultan Khar's family has roots in the village of Khar Gharbi located near Kot Adu in Muzaffargarh District. The Khars have played an active role in national politics since the 1970s: Sultan Khar's younger brother Ghulam Mustafa Khar once served as the Governor of Punjab, and more recently, his niece Hina Rabbani Khar was appointed Pakistan's Foreign Minister. The Khar family has several landholdings including land in Rangla. Around 80 per cent of the land in Rangla is state-owned while the rest is in private hands.

Sultan Khar is the eldest of his family and prefers to live in his ancestral village rather than the cities of Lahore or Islamabad where other members of his family have relocated. He has built a 'dera' (guest-house) near the largest lake in Rangla and is keen on the idea of protecting the wetlands as a game reserve and helping to enforce a ban on hunting in 2001.

'There used to be lots of game here when I was growing up in this area, including gazelles, deer, and partridges. Now, it has all been destroyed. People from outside this region, from Lahore and Multan, have started killing indiscriminately. I used to take an active interest in controlling hunting in Rangla. I was very firm with poachers, but since I developed a heart problem a few years ago, I have stopped going there,' he says as he feeds pigeons at his faded art deco mansion in Kot Adu with its peeling paint and overgrown garden.

The control of hunting is now in the hands of the Punjab Wildlife and Parks Department, and Sultan Khar feels they are not doing an adequate job. 'If I was healthier, I would be sitting there in vigilance and not letting anyone hunt there illegally. Now

Malik Sultan Khar of the influential Khar tribe, pictured here in his guest courtyard, was enlisted by the government to ensure that the Rangla Wetland Complex would be protected from unauthorized hunting.

I fear no one is doing anything. There are barely two government wildlife guards in Rangla.' He says his sons and nephews have no interest in the game reserve and most have moved to the big cities to study or pursue careers. The women of the family in this deeply conservative area are not allowed to move around freely in their district so they cannot become involved with Rangla's management.

Sultan Khar was actively involved in the area from 1988–1997 and government officials agree that he was an effective leader who made a big difference to the management of the reserve. His safeguarding of the lakes protected the wildlife and there was a visible increase in the number of ducks and partridges. However, Khar also used hunting to patronize important politicians. Then President of Pakistan Farooq Ahmad Khan Leghari (1993–1997) visited the area in 1995 with a large hunting party. Thanks to Sultan Khar's brother, Mustafa Khar, who was serving as Minister of Water and Power at the time, electricity was brought to the area in 1995 and piped water was also provided to the local villagers through access to the nearby Wunhar canal.

The local communities living in Rangla are mostly from the Khandoya tribe, who are historically woodcutters and camel herders by profession. They would sell the local wood, transported by their camels, and gradually settled onto the land and became small farmers. The land is not fertile, hence they can only grow enough wheat to feed themselves, supplementing their income by making sacks out of goats' hair and selling them along with the Peelu fruit in the bazaars.

They continue living in traditional mud houses with thatched roofs as if nothing had changed over the centuries. The only indication of the twenty-first century is the presence of electricity poles that serve around two hundred and fifty households in the area.

They hunt in the nearby lakes, despite the fact that it is illegal and they could be fined from three thousand to five thousand rupees. Unlike the city folks who often kill dozens of ducks in one hunt, the locals use primitive guns and can only afford a few

Arched reed silos used by the locals for the storage of grain crops.

cartridges at a time. Hence they only kill one or two ducks in a shoot and use the meat for their family's protein demands. But many of these villagers facilitate hunting by outsiders, who pay them to be their guides and pickers to retrieve dead ducks from the lakes during the winter season.

All the local communities own livestock and depend on the vegetation of the lakes for fodder and livestock grazing. There is no gas supply in any of these villages, so the use of fuelwood as an energy source for cooking and heating purposes is predominant. The majority of the households of the four villages in the vicinity of Rangla use wild sugarcane from the game reserve for making the roofs of their houses.

Since the literacy rate in the area is so low, it is a difficult job to raise awareness about the importance of conserving the wetlands. Many say that the 'marshes are of no use to us' even though they are patently dependent on the wetlands resources. With no schools or hospitals here, the infant mortality rate is very high. If the Rangla wetlands are effectively managed, the locals could be trained to become wildlife guards, giving them a means of employment derived from the protection of the natural resources upon which they depend.

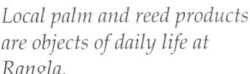
Local palm and reed products are objects of daily life at Rangla.

Tanvira, the huntsman of Rangla Lake, uses his last home-made cartridge to kill a moorhen
for his family dinner. In the autumn, he act as a 'picker' for bird shoots by
urban hunters in exchange for essential hospital care in the city.

Tanvira: The Huntsman of Rangla Lake

At the crack of dawn, twenty-year-old Tanvira and his childhood friend Asghar crouch over a wood-fire in the courtyard of their mud-brick house. They are burning a bush called 'atha' which grows in the desert; they will extract its ash to mix with sulphur and lead shot. This makes a potent home-made *barood* (gunpowder), which they will fill into hollow cartridge shells they have carefully retrieved from last year's hunting season on Rangla Lake. With only two empty cartridges remaining, they know this morning's shoot must be conducted with absolute precision. Tanvira's hand shakes as he mixes the powder and tamps it into the empty cartridge—only last year a hunter from their village was blinded when this home-brewed mixture blew up in his face.

It is the month of May and all is still around the extenuated shape of Rangla Lake. The air is full of the sounds of songbirds and the gentle rustling of the typha grasses that grow in the water. Groups of coots and moorhen are clustered around the reed islands in the lake, searching for their early morning feed. In this stillness, Tanvira crouches in the reeds, his gangly legs shin-high in brackish water and his lime green *lungi* (loin cloth) wrapped up around his waist. His naked torso glistens in the early morning sunlight as he holds the rifle in his hand.

This shotgun is borrowed from Asghar, and is also a home-made contraption with a butt made of coarse mulberry wood. Nearby on dry land, Asghar stands silently, moving only his eyes to track the movement of the ducks. He is the most striking figure in the landscape in his turquoise *shalwar-kameez*. In this poised silence, the two boys hear the distinct sound of a four-wheel vehicle approaching from the south-east. Asghar, the look-out, starts to panic, assuming it is one of the wildlife guards on a rare random check in this wildlife sanctuary. Caught in the act of an unlicensed shoot, they would have to pay a fine of fifty rupees each. Although this is not a large sum, it would nevertheless be a problem for him to part with cash.

Through eyes squinted against the morning glare of sunshine, Asghar looks at the approaching vehicle. Even before the vehicle stops in front of him, Asghar realizes that these are not officials of the wildlife department but the occasional birdwatchers who sometimes come to visit the lake from Multan, which is half a day's drive away. After asking directions to the next lake, they drive on and Tanvira emerges from behind the reeds where he has been hiding.

The sun is coming up and the birds will soon be concealed in the clumps of reeds around the lake. Tanvira spots a shimmering blue male mallard and takes his chance. He aims the rifle and shoots. The mallard rises up in a flurry of bird feather—he has missed his shot. He only has one cartridge left, which will be his last chance until the autumn, when the birds and the seasonal hunters return to Rangla. He takes aim once again and this time he kills a moorhen, which makes for lean and chewy meat, but is better than nothing.

Bird in hand, Asghar and Tanvira slowly walk back to the animal pen where Tanvira's cows and goats have been penned with dry thorn bushes for the night. The animals are shifting impatiently, waiting for their day-long graze in the desert scrub. Tanvira is a shepherd and will soon be taking these animals, which belong to a number of households in the village, for grazing.

After a quick headcount to ensure that all the animals are there, the two boys settle down in the courtyard to cook their moorhen. They make a curry, and talk about better times ahead. In October, when the desert heat cools down, flocks of birds will begin to arrive in their homeland. Although their mind's eye cannot stretch over geographical boundaries, they know the birds come from snowy climes that are in fact the stretches of Siberia. The birds come to Rangla on their way to the wetlands of Sindh, where they spend the entire winter before returning to their northern summer home.

They wonder when Dr Ahmed will be getting in touch with them next—he is an avid hunter who has a thriving practice at Nishtar Hospital in Multan. Asghar's cousin is due to go to Multan with his pregnant wife, who will be examined by Dr Ahmed. In the past few years, Dr Ahmed has been a regular visitor to Rangla. The duck shoot gives him a thrill and a distraction in which he can forget the demands of city life and his extended family. In the winter months, he arrives in his four-wheel drive equipped with good hunting rifles and plentiful cartridge supply. Tanvira and Asghar are his trackers and 'pickers'. They keep an eye out for the new arrivals of birds at the lake and find the best tracks through the marshes, where Dr Ahmed can crouch unseen by the birds to get a good shot. After the shoot, the boys go and retrieve the dead birds that have been shot on the lake.

In return for their services during hunting season, Dr Ahmed comes stocked with rations of food from the city which help tide the family over during the lean months of the agricultural year. Tanvira's mother carefully stores away the tins of cooking oil and bags of sugar to be used sparingly. Dr Ahmed is the only link that Tanvira's family has with a hospital and showing his reference chit in the corridors of Nishtar Medical Hospital gives them special access that would otherwise be denied to them.

The sun is now high up and blazing down into the sand-dunes. Tanvira picks up the clay water pitcher that his sister had filled early in the morning and perches it on his shoulder — the water is all that he will have to last him till sunset, when he comes home after a long, hot day of grazing the animals in the desert sun.

Introduction

I n this chapter, we re-conceive the Cholistan desert as a place of ancient history that has only recently been investigated by archaeologists. This has long been a desert of great aridity and ephemeral fresh water, and its pastoralists are highly adapted to its natural rhythms. To tamper with the desert environment or to erect boundaries and fences is to endanger this delicate social and ecological system. Nothing permanent endures in this desert system that is driven by the movement of water, wind, and sand; here, people and animals must also move in rhyme with nature and respect the rigours of its aridity.

The Ghaggar–Hakra River once flowed through this region, but has now disappeared leaving behind illusory notions of a desert that can once again bloom. But archeological scholarship informs us that this river dried up long before the Harappan Civilization flourished around 2000 BCE, leaving behind a seasonal waterway fed by the Siwalik Hills, comprising a relatively small watershed and reliant on monsoon precipitation that afforded settlements for only a part of the history of Harappa. The great myths of the holy Sarasvati River of the Rig Veda continue to carry their inner meanings in Hindu tradition, even as the outward forms and locations of the Ghaggar–Hakra itself have changed or are lost in the mists of time.

Misunderstandings about a past verdant and river-fed area have led to failures by experts and politicians to alter this desert and its very nature. We explore the Cholistan ecosystem, with an emphasis on a fragile and long-enduring desert and matching pastoralist lifestyle. Indigenous knowledge of freshwater management and understandings of the micro-locations that offer plant materials are described. A small group of national scientists with an understanding of arid zones point out that fencing the desert and altering traditional livestock keeping practices are two of the most destructive interventions that can be imposed on this system. An apt warning that science cannot comprehend all the complexities of nature — even in a desert, which seems a deceptively simple ecosystem.

(facing page)
In Cholistan, a desert of great aridity and ephemeral fresh water, pastoralists are highly adapted to natural rhythms, living with exceeding water stringency.

A series of dramatically beautiful desert forts are strung across Cholistan in various stages of decay, many dating back to the later Mughal era when they served Rajput outposts. But we find that this route through the Greater Rajasthan Desert is a very old one, with layers of history dating back into antiquity, and it is one of the byways of the Silk Route originating in China, opening up a flow of travel and goods West to the Roman Empire and East to India and beyond.

An essay contributed from India's Rajasthan provides a perspective on how people and history are changing in the region adjacent to Pakistan's Cholistan Desert. It refers to the Rajput history and to the present-day Muslim population, where before the fenced border was erected in 1997, families, animals, and wildlife roamed freely as they always had.

We trace the ancient lineage of the pastoralists who populate the Cholistan Desert as warrior-herdsmen who settled here about a thousand years ago. A Hindu Kshatriya tribespeople, they still ascribe their spiritual allegiance to the medieval Sufi saints who inspired conversions to Islam in the eleventh and twelfth centuries CE and whose shrines are visited even today in the nearby cities of Multan and Pakpattan.

One such Cholistani herdsman is Chacha or Uncle Shakir Ali, who is an expert tracker, and our cameo piece focuses on how his knowledge of every track and plant in the desert earned him employment in the household of a United Arab Emirates king who built a palace in the region. He assists the Arab visitors to go deep into Cholistan as they use falcons to hunt in the desert, an ancient part of Arabian chivalric culture.

The final piece in this chapter focuses on the prized winter guest of Cholistan, the Houbara Bustard, an endangered and shy migratory bird that comes here from Central Asia, and all the attention it gains as it forms a valued part of the falcon hunt.

The ancient lineage of the Cholistan pastoralists is traced to warrior-herdsmen who settled here about a thousand years ago.

A series of dramatically beautiful desert forts are strung across Cholistan in various stages of decay, many dating back to the later Mughal era when they served as Rajput outposts; shown here is Derawar Fort.

One wonders if the forts that stand in the middle of Cholistan, extending into
Rajasthan, were strung around near a river that allowed boat traffic.
Paleoarcheology unravels many speculations as misleading.

Ghaggar-Hakra: The River that Was

Until recently, archaeological research has led us to believe that what we see in Cholistan, a wide dry riverbed buried in the sand, is the remaining scar of a once well-watered region that supported agriculture: a river or an inland delta. Folk tradition corroborates this in Cholistan with lore of a 'lost river of the Great Indian Desert'. One wonders if the forts that stand in the middle of Cholistan and extend into Rajasthan were strung around or near a river which allowed boat traffic. Was the ruined market town around Derawar somehow linked to a river? The city of Multan, we know, commanded a central position in western India as a river-based urban capital on the Indus. And where does the sweet water found in some of the deep wells within Cholistan come from? There seems to be no apparent source of water recharge in this barren land; perhaps an old river basin has left behind water deep under the sand.

A branch of paleoarchaeology investigated by an arcane group of scholars unravels much of the above observation as misleading — except the one about sweet water and its links. When boreholes were drilled in the channels of the dry riverbed in nearby Indian study sites, it was found, through a technique called isotope analysis, that the sweet-water that runs under the desert is Paleolithic water; it has been stored there for more than 15,000 years ago by a water channel that once ran through the lands.

Indeed, the recent proliferation of satellite imagery confirms what the eye sees in the Cholistan Desert: a track on the face of the desert roughly parallel to the course of the Indus, but lying south-east to it, starting somewhere in the north in India near the Siwalik hills and ending abruptly in the middle of the Thar Desert of Pakistan.

Thus unravels the mystery of the Ghaggar–Hakra River, the river that once was. The story of this river belongs to the debates on environmental history of the subcontinent in the realm of the greater Indus River system. In a map of 1893, made by major surgeon C. F. Oldham of the Indian Army, who was also an indologist, the Ghaggar–Hakra starts its course in the Siwalik Hills, flows through Haryana, where it can still be seen flowing in a good monsoon year, continues through the Rajasthan Desert into Cholistan and through into the Rann of Kutch, its final destination. At this time in the late nineteenth century, scholars believed that the map confirmed it to be the Sarasvati, the ancient mythological river revered by the Hindus.

Archaeologists have now surmised that although much of Oldham's map is accurate, its paleochannel visible from space, an ancient coastline may have existed millennia earlier further inland to the Rann of Kutch, constituting a paleo-coastline which drained the Ghaggar–Hakra. Found along the channel of the lost river, there are three hundred and sixty scattered settlements dating back to the Indus or Harappan Civilization from around 2500–1900 BCE.

So the main controversy around the river is not really its course, which can now be seen through satellite imagery, but its age and flow of water.

The River Sarasvati from the Rig Veda culture describes it as the 'best of mothers, best of rivers, best of goddesses'. It is regarded mythologically, as the holy river flowing between the Ganges and the Indus around 1500 BCE.

But science suggests that there was no big river here at all, only a seasonal water course affected by the monsoon during the Harappan times. Thus the question arises: when was it an active river that sustained human civilization?

Recent years have allowed new archaeological techniques to date river sediments by measuring the passage of light through quartz deposits found in the sediment. When the youngest sediments, at forty-metre cores, were extracted in the Ghaggar–Hakra bed, it was found that water deposition by the river ceased after about 14,000 BCE, long before the Harappan Civilization.

The second proposition was that it may have been fed by the mighty glaciers of the Indus watershed, which would produce a river like the Indus. This too has been disproven with isotope analysis of mineral deposits. There are no Himalayan glacial deposits in the paleochannels of the Ghaggar–Hakra; it seems

Seashells lie buried mysteriously in the sands. With its paleochannel visible from space, the Ghaggar–Hakra River may have drained millennia ago into a paleo-coastline inland from the Rann of Kutch.

always to be fed by the watershed of the Siwalik Hills, a much smaller source of water.

Having established where the river flowed from, and that it received seasonal and irregular precipitation from the monsoon rains, the question that completes the picture is of how the river dried up.

We know that at the mature stages of the Harappan Civilization this river had already been desiccating for a millennium. The Harappan pottery litter excavated from its middens, or historic rubbish heaps, had been thrown by residents in the already drying river. This desiccation process is thought to have been brought on by tectonic shifts in the earth's geological plates in an area of the Sutlej–Ganga plains that remain tectonically active to this day. These tectonic plates are thought to have undergone an uplift, a folding and creasing up of the land that created a water flow in an incline away from the sea back towards the mountains. In this way, the headwaters of the Ghaggar–Hakra were captured or merged with the Yamuna River, and over geological time the river finally dried out. Where it was once a seasonal river that could sustain some urban settlements, its final drying-up may have become a contributory cause to the deurbanization of the late Harappan period, in which people scattered east and resumed agricultural life.

But what of the myths of the mighty Sarasvati? Michel Danino, who has recently written a book on the river, points out that Indian myths, like all living myths, evolve so as to carry the meaning inherent in them, even as the forms of the myth may change. The *Mahabharata*, coming after the Rig Veda, informs us that the 'she river' disappears and invisibly meets the Ganga and Yamuna at their confluence in the ancient past.

The scholarly controversy surrounding these understandings is centred on whether the Ghaggar–Hakra was important to the Harappan Civilization or not. Andrew Lawler, Science magazine's correspondent on this subject, points to some context in this debate. Our slow understanding of the river's importance is ascribed to a small coterie of paleoarchaeologists spanning Pakistan, India, America, Europe, and Japan. Piecing together a cohesive picture of the river and its history has been hampered by a number of factors: the cross-border hostilities that plague the Indo-Pak subcontinent, the slowness to publish by this group, their reluctance to work together, and the lack of a journal dedicated to this subject have all contributed.

There is a current consensus among scholars that the Ghaggar–Hakra was drying well before 15,000 years but that its desiccation process still offered the Harappans a source of water in addition to the Indus for some centuries of their civilization before they scattered and moved east. Scholars point out that we understand little about how this civilization viewed water other than its instrumental uses in agriculture, town water supply, and trade routes. A frustrating deficit in deciphering the Harappan script has not helped, and scholars grapple with any ancient religious concepts or worldviews about rivers and water. Possehl, a foremost Harappan scholar, suggests that, 'the Indus obsession with baths, wells, and drains reveal a religious ideology based on water as a central [cosmic] theme.' In some ways, this ancient attitude to water rings true to this day in what is now Pakistan.

The ruined market town around Derawar is located on an ancient part of the
Silk Route passing through the Greater Rajasthan Desert.

The Arabic word for a dry-place without water is 'barr'; it is a word used to denote 'the other', somewhere outside the realm of the desirable. Pakistan's Cholistan Desert is one such place, occupying 2.6 million hectares.

The Barr Sands

The Arabic word for a dry place without water is *barr*; it is a word used to denote 'the other', somewhere outside the realm of the desirable. Pakistan's Cholistan Desert is one such place occupying 26,000 square kilometres in the south central extension of the Greater Rajasthan Desert. From realms of wishful thinking and collective fantasy, we act with a shallow understanding towards this desert. Additionally, we tinker with strategies borrowed from science, development, and political expediency which threaten the very existence of this fragile but ancient social and ecological system. Its sand-dunes are dated by archeologists back to some 10,000 years, and its people settled here one thousand years ago in a landscape that seems to have barely changed at all. It is not a system that can be fenced and bent-out of shape to yield the familiarity of settled living and verdant slopes. Instead, we must humbly submit to its laws and respond to the few demands of the pastoralists who live here to enable for them 'an even chance' to continue living in the modern era on their terms without destroying their identity or their *barr* sands.

Recent perspectives reveal a misplaced charity that cannot help the people and desert of Cholistan. Classified as arid rangelands, with a population in poverty who are continually forced to move, this desert is alternatively viewed as a potentially 'pleasant, forested, and friendly environment' by well-intentioned scientists and politicians alike. In addition, the misunderstanding about the Hakra River has helped to fuel these notions of a possible future based on a misconceived past. It is often thought that here the Hakra River nurtured a fertile agricultural civilization; instead, archaeologists have discovered that the tail end of Harappan settlements did coexist on the banks of a seasonal waterway, but as tectonic plates shifted, the water drained away relatively quickly and settlements were abandoned. The Ghaggar–Hakra was never a mighty river like the Indus, and began drying out thousands of years before the Harappan Civilization. We can conclude that it has been a sandy desert for at least a millennium.

Nevertheless, multiple government and international investments have yielded failed experiments to reseed the desert, to irrigate it, to enhance its biomass yields. Scholarship about similar attempts on the fringes of the Sahara and arid highlands of Tibet now reveals how the land and the people elude these attempted improvements.

In the Arid Zone Research Station of the Pakistan Agricultural Research Council in 1996 was a small group of scientists who offered some insightful solutions to this puzzle. They suggested that it is imperative to understand the delicate balance between the living and non-living components that comprise the Cholistan Desert. They recognized that the demands of the Cholistani pastoralists are few: more drinking water and healthcare for humans and animals, and better markets for animals and animal products.

On the other hand, the longer-term demands of the desert are also few: don't erect fences and don't try transforming this ancient desert. With the construction of the India–Pakistan border fence in 1997, we have already broken one of these rules of thumb.

These scientists also explain that aridity is the most enduring feature of Cholistan, with wet and dry years occurring in clusters. The desert is divided into two geomorphologic regions: the northern, or lesser Cholistan borders some canal-irrigated areas made of saline alluvial flats with clayey deposits at only a depth of 30–90 centimetres, and Greater Cholistan, which is wind-blown and sandy comprising of ancient river terraces and large sand-dunes at an average height of 100 metres.

Life is sustained around a fine-tuned and ephemeral system of fresh water collection that must move over time and space. In the clayey areas of the lesser Cholistan, pastoralists make water ponds called *tobas* used for human and animal drinking where the water is fairly clean except when slowly contaminated by the animal waste that accumulates as the season progresses. In this vast open space, *tobas* are common property based on historical clan traditions. The animals, consisting of indigenous breeds of camels, sheep, and goat, graze in concentric rings around the *toba* such that in the rainy season, they are within a one kilometre radius around the *tobas*, and as the monsoon water percolates under the soil surface, they graze in extending rings of up to 15 kilometres radii, following the grasses that spring around the slowly seeping freshwater. During October, when the waters from monsoon are completely depleted in the desert, clans move their herds to semi-permanent dwellings, where traditional *kunds* or water storage tanks, often cement-lined, await them with fresh water.

Scientists note that a wide range of nutritious and drought-tolerant species of indigenous grasses, shrubs, and trees are found in the region. Although these are slow growing, they respond to the bursts of rain that occur, and provide ample feed for animals. With much micro-variation at each site, pastoralists are familiar with each location based on the spectrum of soil, moisture, salinity, and wind. They estimate that the total grazing area of Cholistan is about 23,000 square kilometres, which provides feed for about 140,000 animals for five months annually, and the actual animal population is estimated at some 10,000 animals more than this number. This delicate balance of how many animals to the capacity of grass afforded by the desert has been maintained by the pastoralists over centuries, but fencing off the desert between political boundaries is a large threat to this maintenance; the trade-off is likely to be eased by monetary transactions with urban areas when herders buy fodder, water, and nutrition for the pastoralist economy in exchange for labour, animals, and animal products.

To conclude with the voices of the Arid Zone scientists, they warn of two acts that will rend apart this desert system: fencing, that is seen as a sign of hatred and exclusion by pastoralists whether it comes as part of transnational boundary marking or as a package of so-called improved range management regimes, and the reduction of herd-size of animals, since the prestige of the clan and security in hard times comes to pastoralists not with animal quality but the numbers in the herd. Both these are conundrums for traditional science, as higher productivity and exclusive management are the tenets of developmental growth. Here, clearly, we will have to rethink these development tenets and return to a

In the clayey areas of lesser Cholistan, pastoralists make water ponds called 'tobas' used for human and animal drinking. The water is fairly clean except when slowly contaminated by the animal waste that accumulates as the season progresses.

The delicate balance of the desert has been maintained by pastoralists over centuries as they traditionally matched the number of animals to the capacity of desert grass.

greater respect for the rules of a desert system of arid starkness that conceals complexity and fragility.

Perhaps the best example of the munificent outsider, if that is what experts and politicians aspire to be, is Sheikh Zayd type of assistance to Cholistan, for he is a desert Arab himself. Under his leadership, the desert pastoralists are simply supplied with water and healthcare services as an outright grant in exchange for his presence there, and these are perhaps the greatest good that the pastoralists wish for.

*Pastoralists who live here expect to continue living in the modern era
on their terms without destroying their identity or their barr sands.*

Traditional water tobas are common property based on historical clan traditions. The animals,
indigenous breeds of camels, sheep, and goat, graze in concentric rings around the toba.

Scientists estimate that the Cholistan Desert offers a total grazing area of about 2.3 million hectares which provides feed for about 140,000 animals for five months annually. The actual animal population is thought to exceed this number by about 10,000 animals.

The Cholistan forts were essentially watchtowers, uniformly erected at an even distance of approximately 29 kilometres and built from gypsum and mud.

The Frontier Defenses

In the desert lands of Cholistan lie the remnants of the ancient Harappan Civilization—mounds and crumbling forts, some with archaeological remains dating back to 1000 BCE. Researchers now point out that the earlier desiccation of the Ghaggar–Hakra River still offered the Harappans some source of water for their well-planned cities before they finally moved and resettled in the east. The vast sand-filled tracts of Cholistan contain remains of the earliest Harappan settlements, like Derawar, which is the only habitation that has survived up till now. There has always been a fort at Derawar, say archaeologists, built on layers of history going all the way back to antiquity.

The ancient fort and others like it were probably used to protect the early trade in silk that was carried out by the great caravans of merchants and animals travelling over some of the most inhospitable territory on the face of the earth. The great East-West trade route, known as the 'Silk Route', was established during the Han Dynasty (206 BCE–220 CE) and stretched from Kashgar onwards to the cities of the Persian, Indian, Syrian, and Greek merchants, from where it finally reached the Roman Empire—often through Jewish entrepreneurs.

At Kashgar, there were many choices: some caravans went westwards over the Terek Pass into present day Kyrgyzstan in Central Asia, while others crossed the high Pamirs to the south and went over the Khunjerab Pass and down into present day Pakistan. From there, they chose either the land route crossing the deserts of Cholistan and Thar to the port of Thatta in Sindh or the boat route from different river ports, like Multan. From Thatta and the river ports, there was the onward journey to the Port of Aqaba in present-day Jordan and then onwards to Rome.

The movements of the caravans of camels continued in the Cholistan Desert throughout the centuries, and historians say, the chain of forts that dot the desert today were specially constructed by Rajput rulers with the intent of monitoring these caravans. These forts were essentially watchtowers, uniformly erected at an even distance of approximately 29 kilometres and built from gypsum and mud. For the provision of potable water, specially designed underground water tanks were constructed within the limits of these forts. Some of these forts were rebuilt during the Mughal period (1526–1857 CE).

The entire Cholistan Desert was once under the control of the Rajput rulers from Jaisalmer, and they built the desert forts that exist today, often on earlier mounds. Considered the grandest of these forts after Jaisalmer Fort, the square Derawar Fort rises from the desert sands like a mammoth castle that can be seen from miles away. Each of its forty circular towers has a different design of geometric and floral patterns and the fort is spread over an area of 1500 square metres. In the year 1733, the Nawab of Bahawalpur conquered this region and his family rebuilt Derawar Fort and lived there until 1948, when Bahawalpur State became a part of Pakistan. In 1825, Nawab Bahawal Khan Abbasi constructed an adjoining mosque with cupolas and domes of exquisite marble. The ancestral graveyard of the Abbasis, as the former royal rulers of Bahawalpur were called, is also situated close by.

Near Derawar is a place called Bijnot, where the border with India is just 25 kilometres away. In the nineteenth century, the British colonial historian James Tod authored an exhaustive account of Rajasthan, and in it made a specific reference to the fort at Bijnot, which was first constructed in 757 CE. This fort was under the command and control of Rajput rulers for nearly a thousand years before the Nawab of Bahawalpur conquered it as well. Presently, this fort houses a contingent of the Pakistan Rangers, an elite force for border patrol.

Some of the other major forts in Cholistan are Forts Abbas and Mojgharh, while smaller ones such as Mirgarh, Jamgarh, Khangarh, Nawankot, and Khairgarh were used as observation towers. The walls of these forts were made extraordinarily thick, up to 10 to 12 feet in diametre, to induce a natural air conditioning effect in the desert heat. Many of them are today in a state of neglect and disrepair. Their mud plaster is crumbling into the desert, adding to the relics left behind by older civilizations.

There has always been a fort at Derawar, say archaeologists, built on layers
of history that include this beautiful Mughal mosque. The guards of the
last Nawab of Bahawalpur's family still live in Derawar Fort
and permission has to be taken to enter its fortified doors.

An elegant Mughal period entranceway, now falling into decay, at Derawar.

*Two friends meet in old Derawar bazaar to animate once again the crumbling
thoroughfare that must once have been a bustling place.*

*Shepherds Kamis Khan (foreground) and brother Jani
at the Kuriya Beri village near Kishangar Fort.*

From India: The Rajasthan Route*

Jani Khan looked regal seated on his camel, just like one of the warrior kings of yesteryear. He had travelled several hours through the night from his home of Shahgarh village in western Rajasthan to reach Dhanana village at daybreak. Dhanana is a dusty desert village of three hundred people, and is situated fifteen kilometres from the Pakistan border.

Jani looked every inch a prosperous shepherd, almost like a businessman with his clean, white, *shalwar-kameez* and the spotless white turban. He traded in goat's milk and understood the balance sheet. Jani's mother was born in Dhanana and he was happy to be with his relatives again. How long would he stay? He did not know. 'Maybe two days, maybe ten days!' he laughed, scratching his beard. His nephew Faizu Khan solved his predicament by saying '*Mamu* (uncle) will not go in a hurry. Tonight, we will sacrifice a goat to celebrate.'

In remote Dhanana, which only recently got its first large tube well, putting to rest its water woes, the arrival of visitors is always a reason to celebrate. Of course, it is not as if they do not get visitors. Every day, at least half a dozen shepherds come from villages close by, with bags of grain that they want to mill in the lone grinding machine available in the village.

The machine does not work every day, however. Erratic power supply ensures dependence on a diesel-supported motor. When there is no power, the shepherds just gather under the large *Neem* (*Azadirachta indica*) tree in the village, pull out their hookahs and chat. Abdul, an elderly shepherd, decided to take his camels to the concrete watering hole adjacent to the tube well in the village. The well-being of their animals is of utmost importance to the shepherds. 'We are happy, if they are happy,' Abdul said simply, as he lit a 'beedi' or hand-rolled tobacco, and blew some smoke with an air of contentment.

The border between India and Pakistan holds a lot of significance to the shepherds. Jani has uncles and cousins in the Mirpur

---

*By Rahul Chandrawarkar including all photographs in this essay.

Abdul with his camels at the watering hole in the Dhanana village.

District of Pakistan: 'Till the late nineties, we used to cross the border easily into Pakistan. We used to stay for fifteen days with our relatives and have a good time. They did the same. Now with the double fencing, which came up in 1998, it is impossible to cross the border now. A huge tragedy,' rued Jani, shaking his head.

Right next to the border is Karada village, a Rajput settlement. Pratap Singh, a former soldier, explained the life of his shepherd community. 'Until the government dug a 900-feet-deep tube well last year, we had a tough time collecting drinking water. Our women had to walk several kilometres into the jungle to fetch water from some wells. The two large wells in our village are now used to supply water to our cattle.' According to Pratap, before the border was fenced, their cattle often strayed into Pakistan and they had to walk behind them to get them back. 'Animals do not understand borders. Perhaps that is why they are better off than us,' he pointed out.

The metallic road connecting Karada to Jaisalmer runs through endless sand dunes on either side, punctuated every once in a while by a row of shrubs. There is infinite beauty in the vast landscape. From some five kilometres away, one can sight the illuminated fort. The yellow lights make the fort look very surreal. It is almost as if one is back in the year 1156 CE, when Maharawal Jaisal established the fort and town.

The desert town has a very special place in history. It has witnessed many a bloody battle between the 12th and 14th centuries. It was first Alauddin Khilji (1315 CE) and later Mohammed Bin Tughlaq (1325 CE) who attacked and looted the town. However, the Rajput women did not fall prey to the marauders. They committed *jauhar*, a Rajput tradition of mass self immolation, which saw women jump into fire in batches. History books point out that as many as three *jauhars* have taken place in the history of Jaisalmer: first in 1315 CE; then in 1325 CE; and later during the reign of Maharawal Loonkaran (1528–1550 CE).

Luckily, peace once again returned to the town between the reigns of Maharawal Har Raj (1561–1577 CE) and Maharawal Moolraj II (1761–1819 CE), for the rulers in between decided it was prudent to make peace with the Mughal emperor in Delhi. Jaisalmer was essentially one of the most important trade routes to Afghanistan, Iraq, Iran, and Saudi Arabia in medieval times. It was located on the Silk Route and the rulers of Jaisalmer earned their revenue through the taxes levied on traders passing through the town.

The town of Jaisalmer has grown around its fort, which is different from other historic forts. A tiny town exists inside the fort, and people have been living in it since 1156 CE. Today, the fort comprises *havelis* (traditional mansions), temples and residences, now converted into hotels for tourists. When one walks through the imposing fort gates in the early morning, one is greeted by the pigeons perched on its ramparts. The town itself is built throughout in the local Jaisalmer golden stone, earning it the name of 'Golden City'.

At the entrance of the fort, there is the constant traffic of western tourists. Backpackers are walking in and out of the fort. Tourism has grown exponentially in this desert town. The handicrafts industry has prospered as well, but the rural folk in Jaisalmer District have not benefited from the tourists. In the absence of a strong co-operative movement, village artisans are exploited by middlemen, who make the most of the bounty. Much of the handwoven fabrics come from the neighbouring district of Barmer, another border

A morning view of the famous Jaisalmer Fort with a slightly overcast sky.

Women drinking water in the Dhanana village, 15 kilometres from the Indo-Pak border.

district with Pakistan, while artifacts like metal, brass, and wooden articles come from the northern district of Bikaner.

North of Jaisalmer is the village of Mohangarh — Tar Singh is a shepherd who belongs to this village. Tar is constantly on the road with his flock of sheep for a good five months in the dry season. 'We have no choice,' he said. 'We keep wandering in the entire Jaisalmer District in search of water and grass. We make a bonfire at night and camp in the open. There is no other choice.'

The shepherds are a hardy lot. Tar and his two companions have only a thick shawl made of coarse wool and a cotton blanket to ward off the severe Rajasthan winter. Tar Singh sometimes loses some of his flock in hit-and-run road accidents. 'Most times, jeep drivers do not stop after hitting our sheep. Sometimes, we are able to note down the number and complain to the police. At other times, we are able to stop the driver and demand a compensation of a few thousand rupees,' he says.

Another important fort in the area is Kishangarh, one of the border forts of the former Maharajahs. Kishangarh is a brick fort and its ramparts display large cracks. It is today uninhabited and only visited by shepherds like Chacha Rahim, who lives in Changaniyo-ki-Basti, a shepherd village about 25 kilometres from the border. 'We have to fend for ourselves in the desert, sometimes for months together, when we are out grazing our animals,' explained Chacha. According to Chacha, life has become easier in his village since a tube well was installed by the government. Water from this tube well is stored in a large concrete tank. This water is supplied to animals and villagers alike.

'Before the tube well came, we led the life of nomads completely. We would be wandering in the desert in search of shrubs and water for months on end,' he recalled. 'Life is far easier now. We really do not have to thirst for water anymore. The Indira Gandhi Water Canal was a boon for us.' The Indira Gandhi Canal comprises 650 kilometres of main canal and a 7,750 kilometres long distribution system. The water is transported from the Himalayan mountains into the Rajasthan Desert.

Approximately 40 kilometres from Jaisalmer is the Desert National Park, India's second largest national park. The project was initiated in 1980 with the primary objective of protecting the Great Indian Bustard, which is a protected species and the state bird of Rajasthan. The park is home to a wide variety of wild animals such as the Chinkara (antelope), Desert Fox, Indian Fox, Desert Hare, and Monitor Lizard. Among reptiles, the poisonous Krait and the Viper are the two important snake species, while the park also houses the Sand Boa.

However, the park prides itself primarily for the protection of its bird species. According to the Deputy Conservator of Forests, the population of the Great Indian Bustard has gone up significantly. 'Not only has the bustard prospered in the park, peacocks and partridges have also proliferated.' The park has a work force of hundred and twenty men, many recruited from nearby local villages, who are armed with guns, wireless phones, and camels for patrolling duty. In terms of protection, it is one of the best national parks in the country.

Chacha Rahim (in the foreground), a veteran shepherd from the Changaniyo-ki-Basti village, 25 kilometres from the Indo-Pakistani border, seen here with fellow villagers at the village tube well.

*The people of Cholistan still describe themselves as the Mahars, or Mihirs,
the ancient honorific which means 'of the sun'.*

The Warrior–Herdsmen

The people of the Cholistan Desert are pastoralists, herding sheep, cattle, and camel across green pasturage, moving deep inside the desert when it blooms with winter rains, and coming out into its boundaries in the riverine areas during the hot, dry months.

Their earliest appearance in historical records is in the ninth century, when they are noted as Rajput clans, the 'fire-born' *Suryavanshi Kshatriyas* or sun worshippers, who are thought to originate mostly from the pastoral Gurjaras. There is considerable scholarly debate about who these people really are. Some associate them with the Hun invasions of India in early history; others with a vibrant race of Central Asian tribes; yet others trace the Sanskrit meaning of Gurjaras to cattle breeding, connecting *gau* and *gadar* successively as cow and sheep in Sanskrit. They remain an ethnic tribe spread over Pakistan, Afghanistan, and India, mostly known as the people of the Indo-Gangetic Plain. The British depict them as fierce 'criminal tribes' who participated in the 1857 War of Independence and several of whom were awarded the Victoria Cross for bravery in the battles of World War II. The people of Cholistan still describe themselves as the Mahars, or Mihirs, the ancient honorific which means 'of the sun'.

The world of the Cholistani herdsman until recently represented a figment of the Indo-Islamic world as it was between eleventh and thirteenth centuries when India was conquered by the Muslims. This was 'a world constantly on the move ... characterized by openness and fluidity as it knew no hard and fast boundaries' as described by historian Andre Wink. Their society constituted bands of warrior-herdsmen who were held together by exclusive allegiance to a chosen leader. With nothing to support or perpetuate political institutions, it was a society of mobile wealth of the profits of trade and animal herds revolving mostly around individuals; a loose association characterized by the nomadic societies of Central Asia that seeded the Mongol invasions.

It was also in medieval India that these Rajput clans converted to Islam through two Sufi saints who settled in the western fringes

Thought to originate mostly from the pastoral Gurjaras, the earliest historical records of the Cholistanis is in the ninth century as Rajput clans, the 'fire-born' Suryavanshi Kshatriyas or sun worshippers. Linked to an ancient identity, the sun motif still appears in local embroidery.

of the Greater Rajasthan Desert. They still pledge their allegiance to Sheikh Farid-ad Din Ganj-i-Shakar (d.1265 CE), also known as Baba Farid of Pakpattan, or his contemporary Hazrat Baha-al Haqq Zakariya (d.1263 CE), whose tomb is in Multan. At the time of these conversions, and well into the Mughal era in the late 1500s, these territories were dominated by rural peoples with peasants and nomads comprising 85 per cent of the inhabitants, and townspeople only 15 per cent. Thus the area comprising this rich blend of settled and semi-settled peoples is described as a crucible merged of two types of society—nomadic and sedentary—that underlie both the divide and the dependency between nomadic warriors and settled cultivators. The nomads once represented the military manpower and the mobile wealth that echoes still in the culture of Cholistan's Mahars.

But modern developments like the double barbed-wired fence that divides the borders of India and Pakistan in the Cholistan are unprecedented barriers to the continuity of this cultural history. Families of desert dwellers were often divided with the erection of this barrier in 1997. Within this shrinking world of ecological and cultural confines, we are left wondering about the fate of these warrior-herdsmen in the future; will they rise again as charismatic and vital leaders in the political arena or fade away with their animals like shadows in the falling desert sun?

Conversions to Islam happened through two Sufi saints settled in the western fringes of the Greater Rajasthan Desert, Sheikh Farid-ad Din Ganj-i-Shakar (d.1265 CE), also known as Baba Farid of Pakpattan, and his contemporary Hazrat Baha-al Haqq Zakariya (d.1263 CE) of Multan. Shown here is the muezzin of Derawar Fort.

We are left wondering about the fate of these warrior-herdsmen; will they rise again as vital leaders or fade away with their animals like shadows in the falling desert sun?

A master of the desert, a Cholistani man wears a large white turban used both as a night blanket and a protective shield against the sun by day. Across history, a rich blend of settled and semi-settled people became a crucible merged of two types of society, nomadic and sedentary.

The Desert Tracker

'I'm a jungle bird,' says Mohammed Shakir Ali, his face wizened far beyond his sixty years from the scorching desert sun. Shakir Ali or 'Chacha' (uncle), as he is known affectionately, is an expert tracker who can detect the trails of rare animals like the Chinkara gazelles and birds like the Houbara Bustard in the shifting sands of the desert.

For the past twenty years, he has been in the employment of the Al Nahyan family, the rulers of Abu Dhabi, who own a palace in Rahim Yar Khan. Sheikh Khalifa is the current Emir (ruler) of Abu Dhabi who took over the reins of power when his father Sheikh Zayed bin Sultan Al Nahyan died in 2004. Al Nahyan had a great love for the desert and Chacha remembers the old Emir fondly as his first employer. 'He was looking for a man who knew the desert and I am a master of the jungle,' he recalls, as he unravels his large white turban, which he uses both as a night blanket and a protective shield against the sun by day. Today, Chacha guides the younger Arab princes on their long hunts through the Cholistan Desert in the winter months.

The Arab princes mostly hunt the Houbara Bustard with their falcons, following Chacha into the desert in their convoy of jeeps, transported from the UAE in C–130 cargo planes. Houbara hunting usually takes place in the months of January and February. The Arab princes spend around ten days here. One prince comes and goes and then another arrives. 'They ask me, "Where do we go?" and I guide them. The Houbara Bustard comes to Cholistan in October and we track it and then tell the princes where it is in the desert.'

There are the beautiful and lyrical sights of cranes and gazelles in this desert too, the source of its soulful sung poetry and symbolic imagery. 'We love the cranes; they are a beautiful sight and the desert people won't hunt them or the gazelles. We don't really know where the cranes come from except that they are from distant lands, but we think they go across the border into the hills of India. We are careful about snakes though, especially the vipers; these are dangerous and if we come across them, we do kill them.' He notes the precious herbs of the desert used to cure illnesses like jaundice and hepatitis, and acknowledges that botanists collect many native plants from the area.

For Chacha, the desert, with its undulating sand-dunes and rich wildlife, is home. A member of the Mahar tribe, he was born into a family of shepherds and at just twelve years of age, was told to go into the desert alone with the family goats. 'I was not scared at all, and I did not get lost,' he remembers proudly. From a distance, his father watched as the young boy walked into the desert alone, eyes on the livestock. A young boy's first lonely foray into the desert marks a rite of passage to becoming a man, but is also an economic necessity for the pastoralists who must wander, seeking forage and water for the family's animal wealth. 'We take our livestock to the '*tobas*' (water ponds), which we all have to maintain. There are very few sweet-water tobas; most are salty.' There is much activity at the tobas, especially when it rains and that is where all the shepherds often congregate.

'We learnt about the desert as little children; I was trained by my father and uncles about the ways of the desert.' Now, he says, his own son won't become a shepherd: he has been well educated and wants a different type of a life. Chacha himself now lives in a cemented house in Rahim Yar Khan town and the steady income from his UAE employer means the family need not keep any livestock. 'Those who get educated don't want to tend to the animals. My own son no longer has an aptitude to become a shepherd.'

Chacha was himself born in Bikaner, near Jaisalmer, and after the partition of the subcontinent his family moved to the Bahawalpur District in Pakistan. His maternal uncles and aunts still live in Jaisalmer and before the border fence was built he often went to visit them in their village. There used to be lots of marriages between clans living on the two sides of the Indo-Pak border as well. 'Their desert is the same as ours. We are the same people, divided now by the fence.'

Born in Bikaner, Chacha reminisces the continuity of Cholistan and Rajasthan deserts before the border fence was built between India and Pakistan.

Every autumn season, the Houbara Bustard, a vulnerable migratory bird residing in Central Asia, flies into the warmer climes of Pakistan, resting in the arid shrublands of Cholistan, where it can breed in peace.

Photograph courtesy: Mr Azmatullah, Houbara Foundation International Pakistan

The Prized Winter Guest

Every autumn season, the Houbara Bustard, a vulnerable species of migratory bird residing in Central Asia, flies into the warmer climes of Pakistan, resting in the arid shrublands of Cholistan, where it can breed in peace. A ground-dwelling bird eating mostly desert plants and insects, it used to travel further south, to the Arabian Peninsula. Overhunting of the Houbara in Arabia now means that it no longer travels beyond Pakistan.

Accorded international protection, its hunting and trapping is banned in Pakistan, at least for the locals. However, Arab royalty, mostly from the United Arab Emirates (UAE), have been allowed to hunt the bird in Pakistan in a licence-based system since the 1960s, when Sheikh Zayed bin Sultan Al Nahyan, Emir or Ruler of Abu Dhabi (1971–2004), first started coming to the Cholistan Desert to hunt with his falcons. Presently, the protocols for the Houbara hunt are opaque and the subject of controversy we briefly discuss below.

In Rahim Yar Khan, people recall the 1960s, when a not-so-wealthy Sheikh Zayed used to arrive in a dilapidated train from Karachi at the invitation of the Government of Pakistan. He was drawn to the Cholistan Desert as his sanctuary; away from the affairs of his small emirate of Abu Dhabi. Here he would spend weeks in the desert, travelling during the day for the elaborate falcon and Houbara Bustard hunt, and spending the cool evenings with his entourage under the stars. Immersed in the lyrics of courtly Arabic poetry in the evenings, he relived an essential part of the centuries-old art of desert falconry and its romantic chivalry. Over the years, his love for his personal desert sanctuary converted itself into a full-scale development initiative, financed by Abu Dhabi's oil wealth, in Cholistan.

As Sheikh Zayed's associations with Cholistan deepened, the Abu Dhabi Government funded a number of district-level development projects including a large airport, a state of the art hospital, and the sumptuous campus of the Sheikh Zayed Public School in Rahim Yar Khan town. Included in these grant packages were the construction of metal roads in both town and desert, and an underground water extraction project to provision water ponds scattered in the desert for nomadic communities living deep inside Cholistan. Today, the UAE's Arab princes continue to hunt in Cholistan because they have been granted special hunting permits by the Government of Pakistan.

Yet there has been a dramatic depletion of the Houbara Bustard species in the area due to hunting and poaching, overgrazing of livestock in the desert, and the expansion of agricultural areas. Poachers trap these delicate birds, which are easily frightened, and smuggle them to the Gulf States where they are used to train falcons to hunt them. For modern-day Arabs, falconry is a cultural-symbol, a sport that connects them to their desert roots.

Although Sheikh Zayed died in 2004, his family is willing to continue investing millions of dollars to ensure that the Houbara Bustard thrives, and they continue to hunt with their falcons. Specialized research centres in the UAE study the breeding of the Houbara Bustard and a Houbara Research and Rehabilitation Center in Rahim Yar Khan District in Pakistan reclaims Houbaras taken from poachers to be nursed back to health before being set free. In addition to a laboratory where Houbara diseases are studied, an enclosure with several pens at the centre rehabilitates ailing Houbaras until they are able to fly again. The centre has also conducted research on plants eaten by the Houbara, many of which have dwindled because of expanding agriculture in the desert, and these are being restored through the aerial spraying of seeds in selected areas each year. A similar research and rehabilitation center was also set up in Nag Valley in Balochistan.

According to the environmental journalist Bhagwandas, who is a media champion of the Houbara Bustard,

> *We should understand that hunters come here not out of their love for us but for the pleasure of hunting ... Our decisionmakers should understand that sustainable hunting must follow certain criteria. One, independently collected authentic data regarding the population of birds and status of species and, two, a monitoring mechanism that ensures the bag limit is followed irrespective of the*

hunter's clout. There is no credible data regarding the Houbara Bustards that visit Pakistan each winter and as such it is impossible to quantify how many birds can be culled without imperilling the breeding stock. The ground reality is that when Arab hunters descend on our shores they are awarded the highest protocol possible. Low-level wildlife department staffers cannot even dare go near the hunters' camps and check the bag limit.

Bhagwandas advocates a complete ban on Houbara Bustard hunting for a few years in order to allow their numbers to increase. Wildlife conservationists suggest regulating the hunting process more rigourously, emulating the controlled trophy hunting of Markhor and Ibex that takes place in the north of Pakistan. They point out that if thirty thousand to forty thousand Houbaras migrate to Pakistan annually for six months, then a small number, perhaps a hundred, could be allowed to be hunted in exchange for investment in conservation and community development projects like those initiated during the visionary leadership of Sheikh Zayed.

Photograph courtesy: Mr Azmatullah,
Houbara Foundation International Pakistan.

GLACIERS

SHANDUR

DEOSA

SECTION THREE

The Mountains

Introduction

Deosai is an archetypal wilderness area of the high Himalayas: a large, high-altitude plateau described as a geological hodgepodge of the Karakoram–Himalayas. Its magnificence is both hydrological and ecological. This plateau acts as a vertical sponge, feeding the Indus river system with possibly 20 per cent of its waters owing to its unique position in the folds of the Nanga Parbat Range and the K2, where the monsoon clouds tumble down with their waters as they collide into the high mountains. Ecologically, its high-altitude, extremely short growing season and shallow soils create a marvel of wildflower meadows and the animals that exist there, before disappearing for their long winter hibernations.

In this chapter, we describe the location of this great plateau, its geological marvels, as glaciers have carved out its landscape over millennia, as well as its wetland features. The icy climate and altitude have limited floral types, and have constrained all life forms, small and large, into very special adaptations over time. Its special brown bear population is of particular significance both because of the way the bears live on the plateau as well as the history of how these bears were protected, and how their population increased from less than twenty to more than sixty individuals; showcasing a story that must be valued, told and retold with pride in the country.

Our cameo piece called the last Ibex hunter is a focus on Abdul Wahid of the Astore Valley adjoining Deosai. For him, the hunt is still done in a traditional manner: only for food, with fear, temperance, and wonder in his heart for the grandeur and order of creation. We capture his last hunt, when he knows that the old ways are going forever, and is conscious of new attitudes and new science that values nature in different ways and with different tools.

We relate changing perceptions and actions that impact Deosai's wildlife over a hundred years, reproducing anecdotes of colonial Raj officers who passed through the area en route to

(facing page)
A topographic survey of Lake Saif-ul-Muluk, Hazara, photographed in the 1860s. The British undertook extensive surveys of the mountainous zones which later formed the geographical demarcations for political divisions. (By permission of The British Library, 752/12(141) Hazara.)

*All photographs in this chapter by Javaid A. Khan unless otherwise indicated.

Deosai acts as a vertical sponge, feeding the Indus River with perhaps 20 per cent of its waters owing to its position in the folds of the Nanga Parbat Range and K2, where the monsoon clouds tumble down with their waters as they collide into the high mountains.

Kashmir while noting accounts of bear hunts in their handwritten journals archived in the British Library. Early recognition of its inherent significance earned Deosai a protected status by the Government of Pakistan in the 1970s, but active protection efforts withered away into misinformed paper reports and low budgetary allocations. Then in the 1990s, almost two decades later, a few Pakistani men took it upon themselves to understand and save Deosai in a fascinating story that created a model for high-altitude conservation in Pakistan. The newest efforts resulting from this enhanced importance include management plans and collaborative forums to dedicate effort and resources to conserve this pristine area of value to the nation. Yet persisting threats remain hyper-tourism, roads, dams, and uncontrolled grazing in a land which is so fragile and renders such valuable ecological services that it must surely be left to perform its natural functions without human tampering.

Finally, a rare essay from Tibet by a group working in the Tibetan Autonomous Region of China describes the origins of the Indus in the Chang Tang grasslands of Ngari Prefecture, where it is known as the River from the Lion's mouth. Its sacred association with Mount Kailash and Lake Manasarovar, a wetland thought to be the origin of all the great rivers of the subcontinent, give it a special significance. Yet even in remote western Tibet, overgrazing and the inappropriate fencing of large common grazing grounds is damaging the seemingly boundless grasslands, as are the looming threats posed by highway and rail connections in the planning which may bring destructive heavy machinery used to mine gold, copper, and iron. Even here, in this pristine corner of the earth in Tibet, the ecological integrity and spiritual character of the source of the Indus is undergoing undesirable change. Owing to our extreme dependence on the Indus waters, we must understand and protect all its antecedent sources and tributaries.

Deosai or 'Dev Vasai' means 'Land of Giants' in Shina. The plateau takes its name
from the folk legend of a giant who lived there all year round.

A Land of Giants

Located at an average height of 13,500 feet above sea level, the Deosai Plains are among the highest plateaus in the world. The Deosai Plateau covers an area of almost 3,000 square kilometres in Baltistan. For over half the year, between November and May, Deosai is snowbound; in the summer months when the snow melts, Deosai is accessible by jeep. Situated in Pakistan's Gilgit–Baltistan province, the Deosai Plateau can be accessed via several uncharted ancestral routes. Only two of these have been developed for vehicle traffic: one from the town of Astore in the south-west; and the other from the city of Skardu on the northern side. As has been made clear by recent developments in the region, Deosai's proximity to Kargil, Ladakh, and Kashmir is an important political reality.

Most of the plateau is a common grazing ground for the rural communities that live around it. There are sixteen villages that are settled on the periphery of Deosai, and it is accessible to them for a few months every year during the brief spring and summer periods. Otherwise, it remains a vast and magnificent wilderness, one of the few remaining in Pakistan.

The local people speak Shina, and Deosai or 'Dev Vasai' literally means 'Land of Giants' in Shina. The plateau was named after a famous legend that is narrated by the local people. Centuries ago,

Located at an average height of 13,500 feet above sea level, the Deosai Plains are among the highest plateaus in the world. Only two jeepable routes are developed for vehicle traffic: one from the town of Astore in the south-west, and the other from the city of Skardu on the northern side.

there was a 'Giant' who lived there all year round and grew all the crops he needed for himself on the widespread plains.

The local people themselves depend on agriculture and livestock rearing for their livelihood. Since the region falls in the single cropping zone, most of the villagers grow wheat for domestic use and potato for commercial purposes. They also provide services to tourists visiting Deosai in the summers, as guides and porters.

Straddling the boundary of the Karakoram and the western Himalayas, the Deosai Plateau is surrounded by snow-capped mountains that rise above 5,000 metres. Deosai has now been declared a national park and protected area for the many wildlife species that are found there. The undulating meadows have no trees or large shrubs and come alive in spring when millions of wild flowers begin to bloom all over the lush green grassland.

In an environment that is dominated by intense cold, wind, snow, and ice, with cyclic high levels of ultraviolet radiation, Deosai's vegetation tends to be low-growing and mainly comprises perennial grasses, sedges, and a healthy component of herbaceous flowering plants. Characteristically, many of Deosai's long-lived plants have a bonsai-like adaptation to their icy environment. They grow more roots and rhizome stock below ground level than stunted above-ground aerial shoots, leaves and flowers. These oversized subterranean rhizomes and roots play an important role in water and nutrient absorption and, often contribute directly to the storage of winter carbohydrate stocks, facilitating a rapid response to the advent of the short growing season each year.

A large proportion of small mammal species that hibernate beneath the blanketing snows during the winter rely on these oversized rootstocks for subsistence and, in using them, contribute directly to the dispersal of propagules of the host plant. In all, some three hundred and forty-two species of perennial and annual plants have been identified by botanists investigating the ecology of the Deosai Plateau. Many of these species produce the colourful flowers for which Deosai is famous in the short summer season.

Deosai's summer vistas are mentioned in the diaries of Lieutenant General Henry Lawrence Haughton of the Indian Army in June 1907, where he wrote on his travels in Gilgit–Baltistan. He notes:

> [T]he wild flowers were lovely. Wild irises of many shades from deepest purple to pure white. Great big king cups in many streams and many flowers whose names I do not know. High up where the snow was melted in patches grew multitudes of white and yellow Christmas rose. About sunset, a large herd of goats and sheep thousands in number came down from the pass which they had first crossed. They were the first to cross this year. I gave the men some flour in exchange for some fresh milk...

Once the snow has melted, Deosai appears to be one big wetland criss-crossed with countless streams, springs, and lakes. The largest lake in Deosai is Sheosar Lake, referred to as the 'Blind Lake' in Shina. It is a freshwater lake situated at an altitude of 4,153 metres and is constantly sourced by glacial melt, streams, and spring waters. The lake covers an area of 1.312 square kilometres and supports only three species of fish, one of which, the Snow Trout, is endemic and another, the Brown Trout, is exotic, having been introduced from Scotland during the mid-twentieth century.

As frogs and reptiles are cold-blooded, the icy conditions that prevail for more than half the annual cycle on the Deosai Plains result in a lower density of amphibians and reptiles than that which is encountered at lower elevations in the same approximate latitude. Only one species of frog, the Kashmir Mountain Toad, occurs in Deosai's wetlands. Two forms of lizards, the Himalayan Brown Skink and Glacier Skink, have been recorded from rocky habitat on the plains. The picture with respect to birdlife is very different. At least forty-five resident and migratory species have been recorded, including threatened birds such as the Lammergeier or Bearded Vulture and Himalayan Griffon.

At least sixteen forms of mammals occur seasonally on the Deosai Plains. Species range from the diminutive Himalayan White-toothed Shrew, weighing no more than 5 grams, to the large Himalayan Brown Bear. Males of the latter species may weigh as

Nature's marvels of time and movement have shaped the fabric of rock and water that make up the Deosai Plateau. Signs of ancient glaciation litter the Deosai landscapes in these boulders.

The undulating meadows have no trees or large shrubs. Instead, in spring, million of wild flower bloom all over the lush green grassland.

*Deosai has now been declared a national park and protected area
for the many wildlife species that are found there.*

*The red marmots of Deosai are large, heavy rodents commonly visible on the plateau. They let out loud whistles
of warning as alarm calls when approached by humans and flick their tails before jumping away.*

much as 130 kilograms, which is impressive in the local context but only one-sixth of the weight of males of the Alaskan race of this species, the massive Kodiak Grizzly Bear. Deosai's Brown Bears subsist almost entirely on a vegetarian diet comprising starchy tubers and the rootstocks of sedges that grow on the wet alpine meadows and in the wetlands.

Deosai's Brown Bears hibernate during the long winter months. They typically dig out a large burrow under a suitable rock, usually on a relatively steep hillside a few hundred metres down from the rim of the plateau. The females give birth to their cubs, usually twins, while hibernating in the subterranean den. The cubs are quite small and undeveloped at birth, weighing as little as 450 grams. The families emerge in the spring, once the snows have retreated. An impressive research and conservation programme for the Brown Bears has been established by the Government of Gilgit–Baltistan and the Himalayan Wildlife Foundation.

Tourism is a growing industry of the area, but the large influx of tourists into Deosai in the summer months is starting to pollute the scenic landscape. Overgrazing in the pastures by livestock owned by the local villagers is also a problem to the annual cycles of the flora. As with other great wilderness areas in the world, the management of Deosai National Park has to meet the challenges of solid waste disposal, pollution of waters, and the entry of outsiders into this fragile ecosystem.

On a global scale, the Himalaya, Hindu Kush, and Karakoram Mountain Ranges are relatively young, despite their incredible magnitude. They are, for the most part, composed of ancient sedimentary deposits but also include a significant component of igneous rock, such as granite, and metamorphic rock, which is formed when pre-existing strata are subjected to extremes of pressure and heat. According to tectonic theory, these mountain landscapes have evolved as a result of a continental collision between the Indo-Australian Plate and the Eurasian Tectonic Plate, the former passing under the latter. This drawn-out crash began in the Upper Cretaceous Period, about seventy million years ago, and is still ongoing at a closing rate of approximately 150 millimetres per annum. This is little more than a hand's breadth, but maintained on a geological time scale for over fifty million years, it has acted like a colossal conveyor belt, building up a huge raft of rocky debris and leaving it floating, iceberg-like, on the earth's crust. It is this gigantic geological hodgepodge of igneous, sedimentary, and metamorphic rock that gave rise to the mighty mountain ranges, steep-sided valleys, and high-altitude plateaus like Deosai.

Believed by some geologists to be a western extension of the Tibetan Plateau, Deosai is at once part of and separate from similar landscapes to the east, depending on whether the observation is based on the geology or geography of the subregion. Formed at an elevation of slightly more than 4,100 metres, the plateau is wedged between the Nanga Parbat massif in the western Himalayas and the eastern end of the Karakoram Mountains north of the Indus River. This location has had a profound influence on botanical diversity and zoo geography of the area.

A key component of the process which formed and maintained the Deosai Plateau is the Quaternary glacial and riverine activity and the associated alluvial deposits. This commenced roughly 1.8 million years ago, and still continues today. It is these sporadic bursts of glacial activity that have given shape to and maintained the Deosai Plateau in a colossal process of 'cut-and-fill'.

Signs of this glaciation literally litter the Deosai landscapes. Nearby the Bara Pani campsite lie several, large, irregularly shaped boulders. Closer investigation reveals that the surface of these rocks is scoured with deep, parallel scratches. This is what geologists call an erratic: a large piece of rock that has been carried along for a considerable distance on the surface of a glacier or embedded within it. A little further down the stream, what appears to be a closely packed stone wall stretches for several hundred metres down the length of the broad valley bottom. This feature, which geologists call an esker, was once the rocky bed of a stream embedded within the glacier. In this way, nature's marvels of time and movement have shaped the fabric of rock and water that make up the Deosai Plateau.

An Ibex roams carefree, oblivious to danger from hunters. Traditionally shot only for food,
Ibex are now legally hunted under carefully controlled trophy hunt programmes in Pakistan.

Photograph courtesy: Ghulam Rasool

The Last Ibex Hunter

Rolling onto his elbow where he lay, Abdul Wahid scratched a few large, smouldering coals out of the ashes of the fire. It had smoked a lot, but offered little warmth to him and his two comrades through the long winter night. He leaned forward and gently coaxed the embers into life with a long, steady breath through pursed lips, feeding the newborn fire with half-burnt twigs that lay scattered about on the earthen floor. As the flames flared, he added larger pieces of wood from the dwindling pile behind him. Soon, the fire was popping and spluttering into life, showering sparks about like miniature meteors through the darkness of the hut. 'Juniper', he grumbled under his breath, poor wood to warm oneself by, but there were no options. These past many years, there had been little else to gather for fuel but Juniper scrub and alpine heather in the valley bottoms. The robust forests of his youth were long gone, stripped clean away by a combination of commercial logging and unsustainable subsistence use.

Turning away from the hearth, he nudged open the narrow door. A hand's breadth of new snow covered the trampled ground outside. Dawn was breaking, and far away the summit of Nanga Parbat was ablaze in the light of a new day. *'Al Hamdulillah,'* he murmured. Fair weather meant good tracking: it was a good day for a hunt. Just enough time to offer the morning prayer and wake Shukhratu Khan and Khokaro.

Soon, all three men in the hunting party were gathering their equipment to move out onto the glacier. Shukhratu Khan and Abdul Wahid stooped to tighten the goatskin thongs around their leggings, gently adjusting the insulating layer of moss inside the well-oiled rawhide outer wrapping. The youngster, Khokaro, pulled at the laces of his well-trodden Nike joggers — a legacy of his work as a porter for a German trekking party the previous summer.

They had only one rifle between them: Abdul Wahid's aging 7 millimetre Mauser. Khokaro watched as his father's calloused but practised hands ran over the weapon, working the bolt, checking the rounds in the magazine, adjusting the rear sight. Striking a match, the old man carefully blackened the tip of the rifle's foresight with soot so that it would contrast sharply with his target. Satisfied, he swung the rifle up over his right shoulder by its well-worn sling. 'Let's go,' he said simply, and strode out. 'In the name of God, most mighty, most merciful,' intoned Shukhratu Khan, and turned to follow.

The new snow was underlain by a crisp, crystalline older layer that offered a good footing. By mid-morning, they had worked their way up to the base of the south-west facing slopes that the Ibex favoured in the winter months. As they climbed, they had encountered the tracks of several large male Ibex that had criss-crossed the glacier in search of breeding females during the early hours. This was the best time to hunt Himalayan Ibex, Abdul Wahid explained to his son. For ten days in mid-winter, the males fought furiously amongst themselves for females and carelessly roamed the valley bottoms, apparently oblivious to any danger that might beset them.

Ahead was a steep, blunt-ended spur that projected out from the mountainside to the very edge of the glacier. A relic of a bygone era when the valley had been much colder, this feature had been hacked out of the landscape by a massive river of ice. It would make a perfect place to lie in wait for such a travelling male Ibex. The threesome settled down just below the crest. Concealed in this way, they could easily spot any movement on the opposite sunlit slope.

The early morning wind was still flowing swiftly down the valley like a stream of cold water, and it would be safe to smoke. Sitting with his back to a snowy ridge, Abdul Wahid drew deeply on a hand-made cigarette. His voice low and steady so as not to give their hiding place away, he began to tell the other two of a meeting that he had attended in the adjacent valley just three days before.

A local man from Astore, Muhammad Khadem, had come in the company of a Pathan from Gilgit and one other, to speak with them of matters concerning the Ibex, the Markhor, and the Shapu. 'Shapu?' Khokaro had asked quizzically. 'These are wild sheep that used to live in our valleys when I was young and unmarried like you,' Abdul Wahid had explained. Unlike the Ibex and the Markhor that sustain themselves by eating snow, the Shapu has to drink from unfrozen streams and ponds. This habit confined them to the valley bottoms in winter and made them easy prey, especially for any man armed with a Kalashnikov and blessed with a good eye. 'The Shapu disappeared from our valleys not long after the Russians first made war on Afghanistan,' said Abdul Wahid.

With the Pathan's help, Muhammad Khadem had told them of an organization called 'WWF'. Abdul Wahid enunciated the acronym slowly, for it was a foreign expression that did not roll easily off his tongue. The 'WWF *log*' had made a study of the forests and the wild animals of the alpine highlands, even the Snow Leopards and the great Brown Bears of Deosai. Shukhratu Khan was listening intently. The message that the WWF people brought was that with all these creatures there was trouble. 'No need for foreigners to tell us this creature is trouble,' said Khokaro as he hitched his homespun cape up over his left shoulder with a flourish. 'Did the Snow Leopard not kill so many of our goats and sheep last spring?'

'It is a greater trouble than that,' his father said gently, 'a far greater trouble.' Abdul Wahid paused in his narrative to peer cautiously over the crest of the ridge at the opposite slope—no sign of any Ibex yet. He settled back down on his haunches and continued. Year by year, the WWF people had been counting the numbers of Ibex in the valleys and on the peaks. And they marked that each year the Ibex were growing fewer. The Astori, Muhammad Khadem, confirmed that this was true, for he had himself participated in this informal census. 'Even without counting, in my heart I also know this to be true,' said Abdul Wahid gravely.

'Then it is the will of God!' Shukhratu Khan brushed agitatedly at the crystalline snow clinging to his *shalwar* as he spoke. 'I think it is not so,' Abdul Wahid replied in measured tones, 'I think it may be the fault of our people. Once we were not many, and those among us who owned rifles and were hunters were few. We hunted but with temperance. Now there are too many people and too many guns and far too many hunters.'

After a quick look over the ridgeline, he added, 'The Pathan told also of how the numbers of sheep and goats had increased just as our people in the valleys grew in numbers with each generation. Our four-legged animals and the wild animals are cousins and eat the same food. There is no longer enough pasture for all and so the wild animals grow less with each year that passes.' Soon, like the Shapu, they may be gone forever. 'What need of wild goats and sheep, father?' asked Khokaro, 'Hunting is cold, hard work and already my feet are freezing!'

'The Pathan spoke of a woven carpet,' recounted Abdul Wahid. 'He said that just as there were many patterns on a rug and all were joined by the same thread so God created many forms of life, linked one to another.' The wild Ibex need the pastures to sustain them. The pastures, in turn, need the Ibex to help spread the seeds of the plants, to fertilize them with their droppings, and to loosen the soil with their sharp hooves. If we lose the Ibex, we may begin to lose the pastures also, and with them our livestock.

The one from Astore said that the people in Bar Valley near Hunza had stopped hunting for meat altogether for the past seven years. 'Then what do they do for meat in the winters?' asked Shukhratu Khan. 'Surely, they cannot keep slaughtering their goats?' 'According to their plan,' Abdul Wahid continued, 'the government rewards them each year with a permit for hunting one Ibex.' 'One Ibex!' Shukhratu Khan's lined face mirrored his disbelief. 'It is not the people of Bar who hunt; the permit is sold to a Westerner who pays more than two hundred thousand rupees. One quarter of this money is paid to the government in Gilgit, and the Bar people keep the rest to do as they please. Already it is said, they have built a school for their children with an iron roof and a medical clinic.'

The three men fell silent for a while, and then Abdul Wahid spoke slowly: 'I think that I will ask the other elders of our village to consider this plan. Perhaps they will invite the WWF people to meet with us also.' He turned and slowly raised his eyes above the level of the snow bank on the crest of the ridge. Stark against the snow-white background, a mature male Ibex was picking his path carefully down the opposite slope, not fifty metres away. Without moving his head, Abdul Wahid whispered to the others to lie low, and reached for the rifle. His eyes fixed on his quarry he noiselessly eased the weapon slowly up onto the icy parapet of snow, and slipped off the safety catch. The Ibex had paused and was standing with its forequarters raised on a hummock of snow as it gazed intently out over the valley. Abdul Wahid's grip on his trusty 7mm rifle tightened. Feeling the alpine breeze fresh on his left cheek, he took careful aim on the Ibex's shoulder, then laid off a hand's breadth to the left to accommodate the wind. 'Allah-o-Akbar (God is Great)', he murmured and squeezed the trigger.

Deosai remains a vast and magnificent wilderness, one of the few remaining in Pakistan.

Most of the plateau is a common grazing ground for the rural communities that live on the periphery.

The Brown Bear, considered a keystone species in Deosai, whose status reflects the health of the entire ecosystem, has risen from under twenty individuals on Deosai in the 1990s to more than sixty in 2012, sketching a history to be proud of.

Photograph courtesy: HWF

A Protection to be Proud of

At the turn of the twentieth century, Deosai is recorded in the annals of the British Raj as a place of passage to Kashmir where the daring administrators of the empire would cross Baltistan at Astore in the summer months—pony-trekking over Deosai to Burzil Pass—and onwards into the famed vale of Kashmir. On this male-dominated route, they would perform their feats of the hunt and adventure that was coveted in the Raj as essential to acquiring that peculiarly British trait called 'character'.

Henry L. Haughton of the Indian Army notes in his 1907 private diary that one officer friend 'not being contented with what he had shot [in Kashmir] that is, one markhor, two male Ibex, and a red bear, was going back to Baltistan and Deosai Plain ... and was just pottering about looking for bear.'

A personal account of Haughton's unsuccessful bear shoot at Astore gives us an insight into the backdrop landscape as well as the challenges of this outdoor hunt.

Suddenly, Dost Mohammad spotted a large bear on the far side of the Nullah. He was coming downhill, diagonally across the snow slopes. It was dusk [and so dark] that had he not been on the snow we probably should not have seen him. We sat down and watched him till he entered a patch of birch jungle ... I could only make out an indistinct mass ... and could not make out at first how he was standing. However, I let him have it thinking that if his head was down I should hit him in the skull and if [he stood] up I should hit in the chest. As I shot he reared up on his hind legs, striking out with his fore-paws; he nearly fell over backwards but twisted his body round and landing on all fours made off.

More than half a century later, in the early 1970s, the Government of Pakistan was commended for its interest and commitment in conservation of wildlife by Guy Mountfort who led two World Wildlife Fund (WWF) expeditions to identify areas of some last chances to save unique fauna. At this time, Mountfort recommended that a national park be established at Deosai. This was described as follows:

'Scenically, this is one of the most spectacular regions of Asia, with 26,660 feet peak of Nanga Parbat on one side and the 28,250 feet peak of K2 on the other. It will provide sanctuary for a number of seriously threatened local species, such as the Astore race of the Markhor (*Capra falconeri falconeri*), the Shapu (*Ovis orientalis vignei*), and the Himalayan Ibex (*Capra ibex sibirica*), and also protect the much persecuted Snow Leopard, Himalayan Black Bear and Brown Bear.'

During President Ayub Khan's government, Deosai was notified as a wildlife sanctuary and remained so until much later in the 1990s until it was notified as a National Park in 1993 when a story of two Pakistani nationals, unfolded. We recall this story in some detail based on personal interviews of these pioneers below.

Although Dr Anis ur-Rehman and Vaqar Zakaria had never met before, when mutual friends in Islamabad suggested that they go trekking together in 1987, the two accepted with enthusiasm. Both of them share a passion for trekking, and in the late 1980s the dentist and the engineer decided to go to the Deosai Plains to see its flowers and animals, and particularly the bears that are resident in this remote Himalayan Plateau. Little did they know that their common interest would lead them to high adventure that was to change the course of their lives, and to impact the way the country thinks about its national parks. Today, Dr Anis ur-Rehman owns a thriving dental practice in central Islamabad and close by is the elegant office of Hagler Bailly Pakistan, where Vaqar Zakaria is Managing Director of the well-known consulting firm. Both live double professional lives that sprang from the first trek they were to make together so many years ago.

The natural beauty and impressive grandeur of the Deosai Plains have made it world famous. The plains are home to unique species of plant and animal life, of which the Himalayan Brown Bears are the most famous. Deosai is thought to have created a special habitat where this endemic species has evolved over time.

It was the bears that drew Rehman and Zakaria to Deosai, but their trip proved to be a disappointment. They had heard that there were lots of bears in Deosai, but they didn't see any in four trips there. Out of curiosity the two started to read up and gather information on bears. 'The wildlife department gave us all kinds of numbers, like 625 bears. If there were so many bears, then we started wondering why we hadn't seen any,' explained Zakaria. Mystified by this disparity, they decided to return and carry out further investigations in April 1988. 'Those were the days when we did not know much about bears,' recalled Zakaria, 'we thought that they would come out of their caves and we would be able to film them with our cameras!' That did not happen and the team went back disappointed.

Undaunted, they returned again in October 1992 before the snow had set in and the bears had gone into hibernation. That was when they saw their first bears, a mother and cub. 'It was just pure luck,' Zakaria recalled with excitement. 'We had moved away from areas frequented by people and there they were. When we first saw the bear we thought that it was a yak.' Interestingly, they did not even know what they had seen. They sent their first photographs abroad to specialist groups, and it was through this contact that they learnt that according to scientific records these were highly specialized subspecies of Brown Bears, adapted to Deosai and found nowhere else on earth.

So began a passion that turned into the Himalayan Wildlife Foundation (HWF). Both men wrote to the International Bear Association, who referred them to the London-based World Society for Protection of Animals (WSPA) in 1993. With a grant from WSPA, they undertook the first scientific census of the bears. The facts that emerged about the bears were shocking. In contrast to the official figures, the team discovered that there were only nineteen bears left in Deosai.

Zakaria highlighted the issues that confronted them. 'Here you have a biological resource that appears valuable—it was completely unattended for. There we were, with no management and financial resources, out there feeling as if it had all fallen on

With a grant from an international organization, the Himalayan Wildlife Project
undertook its first scientific census of the bears in 1993.

HIMALAYAN BROWN BEAR PROJECT
DEOSAI PLAINS – BALTISTAN
19 AUGUST—12 OCTOBER, 1993

HALEEM SIDDIQUE	**RASHID QAYYUM**	**RAFIQ RAJPUT**	**VIQAR ZAKARIYA**	**DR. A. RAHMAN**	**NADEEM SIDDIQUE**	**CHRIS MORGAN**
Q. A. Univ.	Q. A. Univ.	Sind Wildlife Board	Project Coordinator	Project Coordinator	Univ. of Karachi	Durham Univ. U.K.

Photograph courtesy: HWF

our shoulders, as we were the only ones who knew what this was all about. We took it upon ourselves in the hope that institutions would take over at some point.' The two friends realized that once the natural resources of Deosai were lost through national neglect and apathy, they would be lost forever.

Their hobby had now become a gigantic task. Threatened by hunting and a shrinking habitat, the bears needed official protection. In the government books, the Deosai Plains figured as alpine grazing areas. With the assistance of an enlightened government Chief Commissioner, they managed to lobby under the Northern Areas Wildlife Act and have Deosai declared a National Park in October 1993. But this was a paper declaration with no map defining park boundaries and no budget to work in the park. Nevertheless, it proved to be the essential link in the chain of events to follow.

Unprecedented in the country's history of wildlife conservation, HWF was appointed an honorary warden of Deosai National Park. This gave Zakaria and Rehman the impetus that the bears had been accorded legal protection, and they decided that come what may, they would spend their summer seasons in Deosai. After extensive dialogues with the local community, HWF realized that the national policy on protected areas that excludes peoples' right to use the area was impracticable in the context of Deosai. Given the traditional patterns of coexistence between people and wildlife, they would have to define their own policies on the ground.

'The people of that area actually have an avid interest in its wildlife,' stated Zakaria. Together with residents, they delineated jeep tracks beyond which no traffic would be allowed. They worked out grazing areas that are adequate for local livestock, and a core zone that is occupied by bears and would remain untouched. They say that they have created practices that are based on mutual understandings between them and the villagers. The key intervention was the establishment of check-posts that monitor all traffic in and out of Deosai. Manned by local residents as well as spare government staff, this soon began to show results in bear protection.

Within five years of conservation work, the bear population increased from nineteen to twenty-eight, and presently it stands at thirty-five. In 1997, the HWF along with the South African Wildlife department radio-collared a couple of bears. Radio emissions from the collars were picked up from satellites and transmitted to computer screens to pinpoint the location and movements of the bears. Using this and other state of the art conservation techniques, HWF studied the little understood habitat, and social and genetic make-up of the Deosai Brown Bear. This work won HWF the prestigious International Rolex award in 1996. The project also received television coverage from South Africa TV and the BBC.

Given that hunting appeared to be a major threat to the bears, HWF's well-known work was soon to become a double-edged sword. Now that people knew where the bears were, hunters had better chances of finding their prey in Deosai. Zakaria recalled, 'By identifying the (wildlife) resource, we actually exposed it to exploitation.' When bear population numbers began to fluctuate between twenty-five and thirty, they dropped everything that they were doing, and set out on a sting operation to find the hunters—it was like looking for a needle in a haystack, give the rugged terrain of 3,000 square kilometres.

They found that the reason for the hunting was to get to bear parts. These are used for medicinal purposes and exported across the border to China. Bear hunting is an old family tradition in the area and most hunters are local villagers. According to Zakaria, a bear would be worth about Rs 100,000. 'We reported sales of bear fat in Gilgit and Skardu to the authorities. But since most of the police were in collusion with the locals they were able to get away.' Over the years, HWF has formed a surveillance network that keeps them and the government informed about the bear trade. Now, Zakaria explained, 'the bear business, we feel, is slowly coming under control.'

Deosai is also a prime hunting ground for falcons as well, an activity that is forbidden in a national park by its very definition. While migrating from Mongolia and Central Asia, falcons land at the plains to rest and feed. HWF discovered a falcon mafia which illegally captured and sold these birds to the Middle East where they are used as part of a hunt. So when HWF came across this mafia, they captured the hunters and handed them over to the police. 'We were naïve at this game,' recalled Zakaria. 'The hunters would bribe their way out and go straight back into business.' The falcon poachers can make as much as one million rupees per bird, which is more than the total budget for Deosai Park. With such big money involved it is no wonder that the falcon mafia means business.

Remote and isolated from the rest of the country, the Deosai Plains remain largely undeveloped for the needs of the local population. It is one of the poorest parts of the country. In the years since they started the Deosai project, Rehman and Zakaria have also tried to help the locals based on the belief that the plains, the wildlife, the people are all linked to one another. 'There are still so many things to do and the government doesn't help much,' lamented Rehman.

Through sheer perseverance, the HWF has been able to get three drinking water projects approved for the local communities. They have also set up health camps and other facilities for the locals. 'This is not our job,' commented Zakaria, 'public money is being used for these schemes, but we have pushed hard to get them approved.' He added, 'We want the government's Wildlife Department to participate seriously as they are the custodians and need to take a more active role.'

Despite all the challenges of saving Deosai, both men feel a visceral attachment to the area and its people. Explained Zakaria, 'We are personally and emotionally very deeply involved with Deosai. Each and every bear means something to us.' Over the years their enthusiasm has infected other people to join them in their efforts.

All their years of experience have turned Zakaria and Rehman into experts in bear conservation, and they are called upon to speak about their experiences at different forums. But the experience has also changed them. 'I was not a wildlife person. Now personally we relate to the wildlife in a very different way. Also, our relationships with the people of the area are very deep, we have become friends with them,' said Zakaria. Rehman agreed, 'It's an intense attachment to the Deosai Plains and the people. It also involves a realization that the whole ecosystem is interrelated. If the bears stay, the park stays. We have to take it as a whole. But it's never going to be easy to save it.'

Now the Directorate of the Deosai National Park is the lead agency for the protection of the plateau with the HWF providing annual survey of key ecological aspects. In January 2011, the Directorate convened a group led by WWF-Pakistan to prepare a management plan for the national park. There is recognition that the area must be connected through corridors of protection to adjoining areas important to the range of area needed for species habitats, primarily the Brown Bear. After this account of protection efforts, Brown Bears, considered a keystone species in Deosai, or a single species whose status reflects the health of the entire ecosystem, has risen from under twenty individuals on Deosai in the 1990s to more than sixty in 2012, sketching a history to be proud of.

In a land so fragile and rendering such valuable ecological services, human tampering must be minimized. Road construction on Deosai's shallow soils is exceptionally disruptive to its ecology.

*The region of Mount Kailash is the origin of several of South Asia's most ecologically and
culturally important rivers, the Indus, the Brahmaputra, Sutlej, and Karnali.
Culturally, it is one of the most important pilgrimage sites for Buddhists,
Hindus, Jains, and followers of Tibet's indigenous Bon religion.*

Photograph courtesy: Jon Miceler

From Tibet: River from the Lion's Mouth, the Source of the Indus*

The Indus River, known as the 'Sengge Khabab' or 'River from the Lion's Mouth' in Tibetan, begins its 3,000 kilometres journey to the Arabian Sea at an elevation of 5,500 metres atop the Tibetan Plateau, where its source lies forty kilometres to the north-east of Mount Kailash. The area is so remote that the source of the mighty river wasn't documented for science until being 'discovered' by the Swedish explorer Sven Hedin in September 1907. The region surrounding Mount Kailash is also the point of origin of several other of South Asia's most ecologically and culturally important rivers, the Brahmaputra, Sutlej, and Karnali, and the location of two of the region's most ecologically and culturally important lakes, the freshwater Lake Manasarovar and the adjacent saline Lake Rakshas Tal.

The uppermost headwaters of the Indus lie in the south-west corner of the Tibetan Plateau's 'Chang Tang' or 'northern steppe' ecoregion. The Chang Tang is a vast region of sparsely vegetated, high-altitude steppe grasslands that covers the entire north-western Tibetan Plateau from the Ladakh region of India to China's south-western Qinghai province, and is dominated by various species of grasses. With an average elevation of 4,500 metres, the Chang Tang has an extremely cold climate and very low pasture productivity. However, in spite of the region's harsh environment, the grasslands of the Indus headwaters region are nevertheless home to a remarkable faunal assemblage that includes such rare and endangered species as the Snow Leopard, Wild Yak, Tibetan Antelope, Tibetan Wild Ass, Tibetan Brown Bear, Blue Sheep, and Tibetan Gazelle.

The Indus headwaters region has been inhabited for centuries by nomadic Tibetan livestock herders, who traditionally resided in yak hair tents and moved their yaks, sheep, goats, and horses seasonally between summer and winter pastures; their high mobility being critical to their survival in the marginal

---

*By John D. Farrington and Dawa Tsering.

environment where pasture conditions can change dramatically from one year to the next. While many residents of the region are still semi-nomadic herders, subsisting in large part on a diet of dried meat, barley flour, and dairy products, the drive to develop western China in recent years has brought many changes to the lifestyle of these nomads, including the introduction of permanent houses and villages, replacement of horses with motorcycles, and the fencing off of privately held pastures which has greatly reduced their seasonal mobility.

Culturally, the Mount Kailash region is one of the most important pilgrimage sites for Buddhists, Hindus, Jains, and followers of Tibet's indigenous Bon religion as well. For Tibetan Buddhists, Mount Kailash (elevation 6,714 metres), or 'Gang Rinpoche' in Tibetan, literally 'the precious snow jewel', is the abode of the Chakrasamvara, a four-faced, twelve-armed deity who is commonly depicted in an embrace with his consort Vajravarahi, symbolizing the blissful union of compassion and wisdom. For Hindus, Mount Kailash is the abode of Lord Shiva, one of the principal Hindu deities along with Brahma and Vishnu in the Trimurti (three deities). Shiva is depicted as the trident wielding destroyer and transformer of the world in the Hindu cosmology, while for Jains Mount Kailash is revered as the site where Rishabhadeva, the mythical founder of Jainism, attained spiritual liberation through Nirvana (spiritual bliss). In the Bon tradition, Mount Kailash is revered as the site upon which the founder of the Bon religion, Shenrab, descended from heaven. Tibetan Buddhists believe that walking the 52-kilometre circuit of Mount Kailash once shall purify the sins of a lifetime, while completing the circuit of the mountain 108 times by foot shall bring enlightenment in the present life. Those who are able attempt to complete the circuit in one day, setting off at four in the morning and returning at nightfall.

Thirty-five kilometres to the south of Mount Kailash lie the shores of Lake Manasarovar (elevation 4,560 metres) and the slightly lower Rakshas Tal, which are connected by a small river, the Gangga Chu. The larger of the two, Lake Manasarovar, has a circumference of 88 kilometres and is the most sacred of Tibet's

In spite of the western Tibet's harsh environment, the grasslands of the Indus headwaters region are home to a remarkable faunal assemblage; shown here are Black-necked Cranes.

Photograph courtesy: John Farrington

Ecologically Lake Manasarovar, Rakshas Tal, and the surrounding wetlands are important as summer nesting grounds for many species of migratory birds; shown here is the Great-crested Grebe.

Photograph courtesy: John Farrington

many lakes. According to ancient Buddhist and Hindu beliefs, the Ganges, Indus, Brahmaputra, and Sutlej were all believed to originate from Lake Manasarovar. Therefore the lake is considered to be sacred to both Hindus, who believe Manasarovar to be the mental creation of the god Brahma, and Buddhists, who believe the lake to be the mythical Lake Anavatapta which lies at the centre of the world, where the Buddha once taught the Dharma (religious code) to his five hundred *arhats* (one who is worthy) while seated atop a lotus sprouting forth from the centre of the lake. At Manasarovar, Hindu pilgrims bathe in the lake's waters, while Buddhist pilgrims circumambulate the lake in a clockwise direction, typically a journey of three days. Although Manasarovar is deeply revered, the nearby Lake Rakshas Tal, which Manasarovar drains into, is known as 'the demon lake' in Tibetan, and is avoided by pilgrims who consider its waters to be poisonous. Ecologically Manasarovar, Rakshas Tal, and the surrounding wetlands are important as summer nesting grounds for many species of migratory birds, including the Black-necked Crane, Bar-headed Goose, Ruddy Shelduck, and Great-crested Grebe.

Mount Kailash and the upper watershed of the Indus River lie entirely within the Tibet Autonomous Region's (TAR) sparsely populated Ngari Prefecture, which, in 2000, had a total population of 77,253, and an average population density of just 0.25 persons per square kilometre. Nevertheless, in spite of the region's low population density, ecological issues affecting the Indus Headwaters-Kailash region are numerous, while conservation initiatives in this extremely remote corner of the Tibetan Plateau are still in their planning stages. At present pressing ecological threats to the region include desertification which is the process of making something a desert, poaching of wildlife, development projects, tourism, and mining.

Perhaps the greatest threat to the Indus Headwaters-Kailash region is that of desertification of the region's fragile grassland ecosystems. Desertification in western Tibet has a number of causes with the largest being overgrazing. While at first look the vast grasslands of western Tibet might seem to be a boundless resource, both human and livestock populations have doubled in the region since the 1960s and continue to grow resulting in severe pasture damage as livestock populations exceed the carrying capacity of the available rangelands. This situation has been further exacerbated by recent pasture privatization and fencing initiatives which have reduced mobility of pastoralists and their herds leading to widespread overgrazing on privately held lands, particularly at winter pastures, where herds may now remain up to ten months of the year. In recent years, the establishment of permanent settlements for livestock herders has also been promoted which may eventually become flashpoints of desertification as herders and their livestock occupy these settlements for longer periods each year.

Another factor contributing to desertification is the over collection of juniper shrubs for firewood, stripping the land of important woody vegetation cover. Traditionally local Tibetan herders used yak and sheep dung for fuel, and primarily used junipers as kindling to start dung fires, since dung burns longer and is easier to gather than juniper brush. However, during the 1960s when administrative centres and permanent towns were established in western Tibet, juniper shrubs quickly became the fuel of choice amongst Chinese migrants who were culturally opposed to cooking their meals with animal dung, the only other fuel alternative in the remote area. According to elderly Tibetans in Ngari, many areas along the upper Indus were once covered in vast juniper shrub lands, however most of the region's juniper cover has now disappeared. Regrettably, at present the harvest of juniper shrubs is unregulated, and junipers in the region are threatened with complete eradication.

A final, uncontrollable factor contributing to desertification in western Tibet is that of global climate change which threatens to radically alter the hydrology of the entire region as western Tibet's glaciers and permafrost melt away. While the full ecological impact of global climate change on western Tibet will not be known for years, annual mean temperatures have risen dramatically and, anecdotally; local herders are already reporting increasing aridity and a large decline in pasture productivity.

To date, the widespread poaching of wildlife has been western Tibet's most high profile ecological issue. Historically, nomads in western Tibet have always hunted the region's large fauna sustainably at a subsistence level. However, the construction of roads in Ngari in the 1950s and 1960s brought a large influx of trucks and high power weapons making commercial scale hunting of the region's wildlife possible for the first time. During the great famine from 1958 to 1961, wild yak and wild ass were widely hunted for meat. However, during the 1980s and 1990s, the primary target of professional poaching rings became the Tibetan Antelope, which was nearly extirpated to supply wool for northern India's lucrative 'shahtoosh' shawl industry. To combat the slaughter of Tibetan Antelope and other wildlife, in 2001 the government banned hunting across the entire Tibetan Plateau and carried out a program to confiscate all privately owned firearms and traps in TAR. However, although Tibetan Antelope numbers are presently increasing, shahtoosh traders continue to operate in the region. With the skin of a single antelope selling for about US$140, there is no shortage of locals willing to participate in this illegal trade. Between 30 November 2007 and 5 January 2008 alone, forestry police in Ngari Prefecture's Gertse County confiscated ninety-three antelope skins from three parties, in spite of the Tibetan Antelope being fully protected by Chinese law.

Traditionally, all but a few Tibetan communities have refrained from eating fish for religious reasons, and today there is a fishing ban in place throughout the TAR to protect Tibet's scantily researched high-altitude fish and ecosystems. However, in recent years illegal fishing on a commercial scale has threatened to wipe out entire fish populations in many of Tibet's lakes.

Rapid development is also impacting the environment of western Tibet, where the construction of large infrastructure and tourism development projects is leading to a brisk influx of migrants, settlers, and tourists to the Indus headwaters region, particularly around Ngari's prefectural capital, Ali, located on

At Mount Kailash, Tibetan Buddhists believe that walking the 52-kilometres circuit once, shall purify the sins of a lifetime, while completing the circuit of the mountain 108 times by foot, shall bring enlightenment in the present life. Many attempt to complete the circuit in one day, setting off at four in the morning and returning at nightfall.

Photograph courtesy: Jon Miceler

the banks of the upper Indus River. In 2006, the largest of these projects were the construction of the Ali Airport, the Shiquanhe Hydropower Dam, and the Ali-Rutok Highway. As part of tourism development in the region, the Mount Kailash area is being promoted as one of Tibet's top tourist destinations, and plans include paving the Lhasa-Ali Highway. While increased tourism will undoubtedly have a positive effect on the region's economy, the environmental and cultural impact of the influx of thousands of domestic and foreign tourists to the Kailash region each year remain to be seen. Increased highway traffic in Ngari will no doubt also adversely affect the migration patterns of a number wild ungulate species, in particular the region's few remaining Tibetan Antelope and wild yaks.

In Tibet, many mountains, lakes, and rivers are considered sacred, and mining these locations was seen as offensive to the deities inhabiting them, which in turn could lead to natural disasters and much suffering for humans and their livestock. Consequently, for centuries there was little in the way of mining in Tibet beyond that done by a few individuals using primitive hand tools. However, this changed with the introduction of heavy machinery to Tibet beginning in the early 1950s, at which time mining became arguably the most ecologically destructive economic activity in Tibet, polluting rivers, destroying grasslands, and contributing to the decline of large ungulates and predators which were heavily poached by opportunistic miners and truck drivers.

In Ngari Prefecture, mining in recent decades was largely limited to small, highly destructive, placer gold mining operations. However in October 2005, placer gold mining was banned throughout the TAR due to the widespread environmental damage and low economic returns resulting from these operations. Nevertheless, two large iron and copper mines continue to operate beside the Gar Tsangpo River, the largest tributary of the upper Indus in Tibet, which may result in long-term environmental damage if improperly managed. With the opening of the Qinghai-Tibet Railroad and the planned improvement of the Lhasa-Ngari

Highway, mining companies will be able to transport mineral ore from western Tibet to markets in eastern China on a previously undreamt of scale.

Thus the Indus headwaters region is not only the source of a number of Asia's great rivers, but is also important for the region's pilgrimage sites, traditional nomadic livestock herding culture, and its populations of rare and endangered large fauna. In recognition of Lake Manasarovar's ecological importance, the lake and surrounding wetlands were listed as a Ramsar Wetland of International Importance in 2004, while in 2005 the Government of China established the 3,100 square kilometres Manasarovar National Forest Park, which includes not only Lake Manasarovar, but also Lake Rakshas Tal and Mount Kailash. However, as the twenty-first century unfolds, with the construction of highways, railroads, and airports in central and western Tibet, the extreme isolation that has preserved the upper Indus region's unique culture, ecosystems, and wildlife is quickly disappearing as the modern world edges ever closer to the sacred Mount Kailash. The long-term impacts of these developments on the ecology and culture of the region remain unknown but will without question bring great changes to the region in the coming years. It is imperative that action be taken immediately to preserve the region's ecological integrity and unique cultural and spiritual character.

Introduction

Owing to their remoteness from centres of production and economic growth, the mountains are often on the periphery of collective imaginations. But Shandur is a mountain pass that creates fever in the blood of those who go there and those who merely imagine the annual polo festival. It conjures images of fast-paced contest, handsome, virile men and highly strung horses competing against a spectacular natural setting of pristine beauty—it seems an audacious joust of men with nature itself.

Known as the world's highest polo ground, the festival happens for a fortnight bringing a melee of people and activity, music, tents, and bazaars that disappear to leave this delicate lake complex and peat land alone for the rest of the year; relinquishing it to seven villages to use it as summer grazing for their animals. What does the presence of princes, presidents, and holidaymakers for those short days leaves behind in this high-altitude landscape?

When researchers measure the impact of the polo festival on this high-altitude national park, they find its ecological health is held in balance by thresholds that are unknown. Yet, the use and abuse of this delicate lake severely tests the limits of regenerating the aesthetic and pristine beauty of the lake and the plateau pass as a whole. Because all who come here prize the beauty of the landscape also, the careless disposal of rubbish, intrusive use of the blue lake water, and the hedonism of the short holidaymakers can be controlled. The trouble is that the pass lies on the border of two separate provinces and governmental jurisdictions that are divided on an imaginary line that runs through Shandur itself. Indeed, the local communities too have relinquished traditional pasture management as now this land is no longer theirs. Used by many, and cared for by none, new efforts are underway to bring the administration and volunteers at Shandur together to care for this wilderness area and to preserve its fragile beauty.

(facing page)
Shandur is a mountain pass that creates fever in the blood of those who go there and those who merely imagine the annual polo festival.
(Photograph: Ayesha Vellani)

---

*All photographs in this chapter courtesy Ghulam Rasool, by permission of WWF-Pakistan and Pakistan Wetlands Project, unless otherwise indicated.

Birds of passage used to stop for rest and reprieve at Shandur on their migrations from Siberia in the spring and autumn, but the rare ones like the Ruddy Shelduck, who were once thought to breed their clutches of ducklings here, no longer stay at Shandur. Instead, they pass onwards to Wakhan through Boroghil Pass beyond which the wilderness truly belongs to them.

An essay from adjoining Afghanistan's Wakhan corridor reminds us of the sources of much that is evident at Shandur. *Buzkashi*, the game of pastoralist horsemen from the northern steppes of the Hindu Kush, is still played there; it is an even more daring contest than the anglicized polo instituted by the Raj. Instead of polo's object of a wooden ball, a goat carcass is the coveted prize as it is carried off and vied for by competing teams. Underneath the flight of birds, here too are the migratory humans, the *koochis*, who still traverse the mountain passes now bounded by political borders, still wandering in the timeless search for green pastures to nourish their animals. This mountain wilderness remains a test of human robustness and character, a haven for wildlife, and refreshment for the sore sights of city dwellers yearning for an archetypal contest in nature's ever present spectacular canvas.

The Shandur Polo Festival conjures a fast-paced contest, handsome, virile men and highly strung horses competing against a spectacular natural setting of pristine beauty – it seems an audacious joust of men with nature itself.

*Little local girls dressed up and excited. We ask: who will care for this
wilderness area to preserve its fragile beauty?*

In the month of July, the celebrated Shandur Polo Festival takes place between teams from Chitral and Gilgit, and this isolated wilderness undergoes a dramatic change.

Polo on the World's Highest Ground

For most of the year the Shandur Pass, the most famous of the high mountain passes in the Hindu Kush Mountain Range, located on the border of Chitral with Ghizar District in Gilgit remains desolate. Then, in the month of July, the celebrated Shandur Polo Festival takes place between teams from Chitral and Gilgit, and this isolated wilderness undergoes a dramatic change. Attracting up to fifteen thousand visitors each year, the festival has become important for Chitral's struggling economy. Polo is a passion for the local people and they take it very seriously indeed. Thousands come from all over the northern mountain region to watch the matches every year, often trekking for days over the high mountains to tent in Shandur for the festival.

In the early 1980s, the Chitral Scouts constructed a permanent rest house overlooking Shandur Lake and a polo ground just below the rest house. From 1986 onwards, the polo tournament became an annual fixture. Visitors are allowed to put up their

A member of the Chitral scouts regiment. The pheasant feather is worn as a striking adornment atop their traditional hand-woven woolen cap.

tents anywhere away from the lake and polo ground. Although the district administration is ostensibly responsible for the festival, the Chitral Scouts end up doing most of the organizing, and they are in charge of the upkeep of the polo ground and the stands which are located just in front of their rest house.

Once the dates of the polo tournament are finalized, a tent city springs up literally overnight with a sprawling main bazaar selling everything from kebabs to alarm clocks. Tented restaurants, hotels and even barbershops are set up, and large generators ensure a steady supply of electricity. During the evenings, there are dances and plenty of fermented drinks locally brewed from apricots and mulberries. The end of a game of polo is always heralded by a dance. The spectators form a ring inside which the musicians are seated with their drums and clarinets made of apricot wood. A volunteer usually steps up and starts to dance. As soon as one dancer tires, fresh dancers are always ready to enter the circle and often hours pass by before it all ends. The graceful, preening steps of the dance are traditional, passed down from one generation to another.

Polo has been played on the Shandur Pass for centuries, although not on the commercial scale that one can witness today. In those days, the polo ground up in Shandur was used mainly for the adjacent villages. The village of Sor Laspur with around seven hundred people is located just below the pass on the Chitral side. However, Shandur's location, a hundred miles from Chitral and a hundred and thirty miles from Gilgit, made it an ideal common ground for teams from these two big towns. There are two polo grounds in Shandur; the official Shandur Pass polo ground near the lake is called Fairy Meadows or *Joot* by the locals, while the practice ground located further down the pass in Khuwar language is called *Mass Junali* (moonlit polo ground). It is said that the British officers often played polo under the moonlight, up in Shandur.

It is said that the British cavalry officers initiated the sport of polo in Chitral, and some historians say that it is from there that they exported it to the rest of the world. The polo that is played up

*There is a famous British inscription about polo in Chitral which states: 'Let others play
at other games … the game of kings is still the king of games'.*

in Shandur is cruder and more vigorous than its anglicized version. Some enthusiasts believe the game has been derived from the Afghan sport of *buzkashi*. Often described as 'hockey on horses', the polo that is played in this region has no rules. There are six members on each team who mount mostly Afghan-bred mountain horses; one match can last for two of the twenty-five minutes long *chukkers* (rounds).

Shandur is considered to be the highest polo ground in the world. For the tournament, horses have to be brought up slowly in stages so that they are acclimatized to the high altitude of 12,205 feet. It frequently takes weeks for the players and horses to reach the pass from Gilgit and Chitral. Often the first to arrive, they erect their separate tented villages a fortnight in advance to allow time to adjust and practice before the festival begins. Polo horses are treated with special care, for players can only use one horse for each match, and should something happen to the horse, they have no choice but to sit out the game. Should this happen, the opposing player also has to withdraw from the game.

Once it starts, the game has no referee to moderate the *chukkar*; what follows is fast paced and furious. The polo ground in Shandur, smaller than those in the plains, is surrounded by a low, stone wall. Players are allowed to block shots and even hit one another yet surprisingly both the players and horses rarely get injured. The players say that because there is no referee they are careful not to cross the line since they know it can cost them their lives.

The final match between the best teams from Chitral and Gilgit is a tense affair. In the early days, Gilgit used to win most matches, but now the teams are evenly contested. Spectators' emotions run high as they watch the game with bated breath; they can be quite unforgiving when their side loses. Before the final tournament, plenty of intrigue buzzes around the grounds in the two rival camps. Overcome with anticipation, spectators often insist on sitting on the stone walls around the polo ground perching dangerously close to wooden mallets, panting horses, and the flying ball.

Despite its widespread popularity with the advent of jeeps, this exciting game is starting to dwindle as horsemanship becomes less central to local culture. Local people don't need to ride horses anymore, and hence there are fewer skilled players. But being a polo player is still to be looked upon as a hero and be idolized by the young. The clattering of horse hooves on the way to polo practice begins in the springtime in Gilgit and Chitral towns, and all look upon the mounted men with their patrician profiles, windswept hair, and fit bodies with admiration and recognition. There is a famous British inscription about polo in Chitral which states: 'Let others play at other games ... the game of kings is still the king of games'.

In July 1986, then President of Pakistan, General Zia ul-Haq travelled up to Shandur to inaugurate the 'Shandur Hut' and attend the first President's Polo Cup Tournament. From then on, the festival has been more or less an annual event. Every year, on the day of the final match between Gilgit and Chitral's A-Teams, the head of state usually flies in on a helicopter escorted by top government functionaries. A special landing area for the helicopters has been marked out on the shores of Shandur Lake and VIP accommodation is organized for them.

All this activity has had an adverse effect on the Shandur Lake as the festival is not sensitively managed. During the tournament, announcements are periodically made urging the spectators to respect the environment and keep Shandur clean, but there is no one to enforce this and people can be seen littering the small streams that are found in abundance on the pass. Before the festival commences, Shandur Lake appears to be pristine with no visible evidence of solid waste such as bottles and cans that must have been left over from the previous year. There is a reason for this: there are many shepherds who live in Shandur during the summer months and they pick up any trash that is left behind after the festival ends. Shandur serves as the summer home for old women and young children from the local villages of Laspur district who use the communal pastureland of the pass for grazing their animals. There are several shepherd camps located on the slopes of the Shandur pass, away from the lakes where they live in stone houses built into the slopes for protection against the strong, cold winds that often blow down the pass. In Shandur, they have plenty of green pastures and glacier-melted water in the streams for their cows, yaks, and goats. During the exciting matches, all the shepherds trek down from the slopes to watch the polo matches.

Just after the tournament ends, a walk around the lake reveals plastic water bottles and juice boxes, tin cans and plastic bags littering the edges of the Shandur Lake. One can also see the many streams that feed Shandur Lake and some come from the direction of the tented city. The people tenting in Shandur don't use water from the lake for drinking, preferring instead to use the plentiful clean, glacier-fed streams coming down the slopes of the pass.

By sunset of the last day of the festival, the large tented city disappears in a whirlwind of dust that hangs over the pass for the entire day. In recent years, more permanent hotels and buildings have been constructed on the pass. A long line of jeeps and people can be seen making their way out of Shandur in both directions, towards Chitral and in the opposite direction towards Gilgit. Indeed, people start leaving immediately after the prize distribution ceremony is over — with many carrying their sleeping bags and tents on their backs. The winning side celebrates on their way home and no one stays around once the matches are over except for the exhausted players who have to bring down their horses in the same way that they were transported up.

The losing team in particular makes sure they leave only when it gets dark to avoid the taunts that are sure to be directed towards them by their disgruntled fans. Polo, for these tough mountain people, is a serious affair, and Chitral and Gilgit have always been bitter rivals. By nightfall, Shandur Pass is once again a quiet place and one can hear the cries of the few birds left on Shandur Lake. Until next year, the birds of passage finally have the lake all to themselves.

At Shandur Pass, a tent city springs up overnight during Polo season with a sprawling main bazaar selling everything, from kebabs, embroidery to alarm clocks.

The late Colonel Khushwaqt ul-Mulk, son of the last royal Katur 'Mehtar' or ruler of Chitral, Sir Shuja ul-Mulk. In 1913, the British gave Mastuj to Shuja ul-Mulk, who passed it on to his son. Debonair and over ninety years old, he lived in the old fort at Mastuj and owned much land around Shandur.

Photography: Rina Saeed Khan

The Last Prince of Mastuj

One hears of stories and poems written about polo in Shandur going back many years, but a tournament started to be held periodically from 1925 onwards, recalls Colonel Khushwaqt ul-Mulk,* dressed elegantly in a tweed jacket as he sits in his garden. He is a son of the last royal Katur 'Mehtar' (ruler) of Chitral, Sir Shuja ul-Mulk. In 1913, the British gave Mastuj to Shuja ul-Mulk, who passed it on to his son Khushwaqt ul-Mulk. Over ninety years old, the English-educated Colonel Khushwaqt still lives in an old fort in Mastuj and owns much of the land in Shandur. The locals still hold him in high regard as the last Prince of Mastuj.

In his youth, he recalls seeing many migratory birds on Shandur Lake and remembers bird shooting there. His family has their own private camping area on the Shandur Pass, located on a high ridge overlooking a second, smaller lake. He feels that Shandur Lake is shrinking because the glaciers in the area are drying up. 'Previously some of these glaciers would come right up to the road, but they have receded over the years. The water level in Shandur Lake is getting lower.'

A keen polo player himself, he recalls, 'I remember I played in a big tournament held in Shandur in 1959. There must have been over 1,000 people there. Polo owes a lot to the British; it was really the British officers who revived the game and made it even more popular. The British turned it into a serious contest.' According to Colonel Khushwaqt, 'Polo first came to Chitral from Iran. In the thirteenth century, there was a big influx of refugees from Iran due to the Mongol invasion of Persia. The persecuted Ismaili population left Persia and came to the inaccessible mountain areas of this region where they settled down. With them, they brought both Islam and polo.'

Colonel Khushwaqt's ancestors originally came from Herat in Afghanistan and married into the ruling family of Chitral. The son from that marriage went on to found the line of the Katur

---

* Colonel Khushwaqt ul-Mulk died on 12 February 2010 at the age of ninety-six.

'Mehtars' of Chitral. In 1950, Chitral ceased to be a royal state and acceded to become part of Pakistan, and by 1969 it had became a district of Khyber Pakhtunkhwa province. Although he has a house in Peshawar, Colonel Khushwaqt prefers to live in Mastuj Fort, which is located on the way to the Shandur Pass. His eldest son Siraj ul-Mulk, who runs an elegant hotel in Chitral, has built wooden chalets in the large garden inside the fort, under the towering maple trees. They get lots of visitors around the time of the Shandur Polo Tournament and remain icons of a cultural past that is part of Chitral's identity.

Mastuj Fort's walls are massive, but the building inside was dismantled in the 1960s when the then President Zulfikar Ali Bhutto threatened to take over all the forts and palaces of the royal families in the country. Colonel Khushwaqt still remembers what life was like when his father was the 'Mehtar', and they were living in the massive Chitral Fort. 'I was sent to live in Mastuj when I was two years old,' he recalls. 'But whenever I visited Chitral I would stay in the fort. It was a good life there.' Colonel Khushwaqt is the oldest, high ranking officer of the British-Indian Army who is still living. He left the army when his father died suddenly in the 1950s. He says that his father 'was a hard man', but that he has fond memories of him—like the times his father would visit him in India where he was sent to study at boarding school in Dehradun.

'I remember meeting him in Delhi and attending a polo match where I had to act as a translator between my father and the Viceroy of India, Wellington,' recalls Colonel Khushwaqt, who has two surviving brothers also in their nineties. One lives in Ayun near the Kalash Valleys and the other lives in Garam Chasma where there are volcanic hot springs, popular with tourists. He had many other stepbrothers for his father had eleven wives. Colonel Khushwaqt and his brothers who are all from the same wife are the ones still surviving.

Colonel Khushwaqt loves to socialize with the guests staying at the chalets, and never misses the annual Shandur Polo Tournament. His second son, Sikander ul-Mulk, is the captain of the Chitral team and they have been on a winning streak for the past few years.

Sikander is in his early fifties but still plays the tough mountain polo. Since his father lost his second wife a few years ago, Sikander lives with him in Mastuj Fort. His father's first wife was a cousin to whom he was betrothed when he was only two years old.

Prince Sikander began playing polo when he was only eleven years old. He remembers when the Shandur tournament began to receive official patronage from the Government of Pakistan. 'In 1982, we had arranged a match between Mastuj and Ghizar up in Shandur. Around fifty people came to watch … it was quite a small event.' Shortly after that match, the Chitral Scouts took over the polo ground and built their three-room rest house—and the polo tournament became an annual affair.

Ecological research reveals this lake is naturally sensitive to external disturbance. At an elevation of some 12,200 feet, it has a small catchment, and was declared a national park in 1993 classified under IUCN's Class II areas of biological significance.

Photograph: Ayesha Vellani

Lake of Unknown Thresholds

Natural systems have high resilience to change, but once a threshold is reached, rapid change can ensue; when one dwells on its meanings, this is a haunting statement of ecological insight. First of all, humans too are part of natural systems, for where else would we belong if not on earth which itself is a natural system. And then the idea of thresholds arises pointing to an end-point beyond which rapid change happens, implying an unforeseen and unpredictable shift that may not be desirable like hurtling from a comfortable room into another whose qualities are not known. If we do not understand and value parts of nature, then how can we estimate thresholds, and how can we anticipate changes that may be triggered by our own actions?

This metaphor is often applied to high-altitude lakes in the Himalayan–Hindu Kush–Karakoram Mountain Ranges. The pre-eminent example of an iconic lake whose value has been damaged by human overuse is the Dal Lake in Srinagar, Kashmir. Since the Mughal era, the lake has been a source of aesthetic and cultural celebration, but in the past fifty years Dal Lake has lost much of its recreational value because of the way its houseboats and tourism has damaged its ecology. It is now reported that it 'literally stinks' which is reflected in the quantifiable decline of annual visitors and setbacks to the local economy.

With the polo festival taking place annually at Shandur Lake, attracting fifteen thousand visitors for over two weeks, the central question is what is the cost of human interaction, specifically tourism, on this fragile and popular place?

Ecological research reveals this lake is naturally sensitive to external disturbance. At an elevation of some 12,200 feet, it has a small catchment of some 61 square kilometres. The wetland complex is composed of two lakes and three streams, with great seasonal fluctuation in length. The maximum depth of Shandur Lake is 3.1 metres or as much as a swimming pool, with little inlet of freshwater through one of the streams. It has no major outlet, and hardly recharges from groundwater or any nearby springs.

Thus water collects in this shallow lake with rain and glacial surface runoff, and the intense summer sun evaporates its waters substantially. This means that concentrations of anything put into the lake can magnify their potency in a matter of days. Important buffering effects to this harsh layout are the reed beds on 10 per cent of the lake perimeter and the peat lands that surround it which act as a neutralizing sponge on its embankments.

Shandur was declared a national park in 1993 and is classified under International Union for Conservation of Nature's (IUCN) Class II areas of biological significance: this designation calls for the protection of ecological integrity and provision of recreational and visitor opportunities in a site which must be maintained in a natural or near-natural state. Indeed, research on Shandur shows that visitors place highest valuation on its aesthetic and personal enjoyment aspects, and locals value its grazing resources for animals. So its IUCN classification as a protected area fits its perceived value quite well.

Related to the recreational activities that happen in Shandur is another far-sighted warning from conservationists: if unsustainable tourism practices are put in place, conservation staff who try to change them will face powerful political opposition. In fact, Shandur is in a state of political and legal limbo because its location is on a provincial border of two separate regions of Pakistan; it is situated on a line that separates Chitral District in

Important buffering effects to the stark layout of the lake are the reed beds on 10 per cent of the lake perimeter and the peat lands that surround it.

Photograph: Rina Saeed Khan

Local graziers have withdrawn their traditional management regimes so that Shandur's alpine pastures are an open access resource used by all but managed by none.

Photograph: Rina Saeed Khan

Khyber Pakhtunkhwa province from the Gilgit-Baltistan region. This makes it particularly vulnerable to an unregulated growth of touristic activities, as it is unclear which province governs it.

Even the local communities, of which there are seven surrounding villages, have withdrawn their traditional management regimes so that Shandur's alpine pastures are an open access resource used by all but managed by none. Traditionally, pastures with clear ownership rights are subject to a practice known as *sakhq* which is a collective system of rotational grazing of the pasturelands so as to allow natural regeneration of the flora of the area; this is still practiced in nearby Laspur Valley. These rules collapsed some fifty years ago when polo activity began to dominate grazier rights to Shandur, and to this day herders remain unanimously opposed to the use of Shandur for the polo festival. Yet, their domestic animals deposit some two hundred times the volume of excrement around the lake as do the humans from the festival.

Problems facing Shandur include the extensive solid waste that is an outcome of the festival, its impact on the water quality of the lake as the multifarious activities of washing, boating, and driving near the lake are intensive even if they happen for a short time. The effects of animal grazing and the deteriorating aesthetic attributes of the lake including the green slime around it and the dust clouds that gather from traffic are problems that have now been studied.

Given that the lake does not naturally flush out or remove these pollutants into its water and concentrates them quickly as it dries out in the summer, the Shandur Polo Festival threatens the very qualities of the water body that it depends on most.

Even though we do not yet understand the biophysical thresholds of the lake, recent efforts have been enacted to protect the lake before it is too late. Following a research thesis by a postgraduate student from Oxford University, the 'Save Shandur Initiative' has regularly been taking place as part of a multi-institutional collaboration. For the first time in the history of the festival in 2008 the Environmental Protection Agency of the region, government departments, the army, NGOs, and community-based organizations began coordinating campaigns to clean up the solid waste left behind in Shandur during and after the festival. Volunteers including tourists urge visitors to use garbage bins and canvas bags when disposing of rubbish, and a landfill site near Shandur has been created for safe disposal as well as analysis of the amounts and types of waste generated each year. There are plans to create a buffer zone, and limit proper campsites at a distance from the lake and important streams, but this remains a race against time, as tourist activities rapidly proliferate.

As the last Prince of the area, Colonel Khushwaqt ul-Mulk pointed out the lake is an essential part of the bigger picture: a staging ground for migratory birds, a corridor for rare animals such as the Snow Leopard, Brown Bear, wolves, lynx, fox, and Himalayan Ibex. Additionally, people do realize that if the waters of the lake are damaged forever, then the eagerly awaited Shandur Polo Festival might suffer a similar fate.

For centuries, Shandur Lake served as an important high-altitude wetland, a safe place for rare ducks like the Gargany Teal and the Ruddy Shelduck.

No Safe Passage over Shandur

The top of Shandur Pass is a picturesque plateau with two small lakes, ringed by snow-capped peaks. The pass is approximately five miles in breadth and perfectly level. There is no surface drainage from the lakes and they are fed by melted snow from the peaks that rise to a height of 2,000 feet above the pass. Over the years the impact of the annual polo tournament on Shandur's fragile natural environment has extracted a heavy price although the lake was declared a national park in 1993.

Before the tournament became a regular event in the mid-1980s, Shandur Lake, which lies in a depression not more than fifteen feet deep, two and a half kilometres long, and half a kilometre wide, was an important refuge for migratory birds. All of Chitral District forms part of a major migration route for ducks in this region. Many varieties are seen in Chitral, and they arrive in February–March from warm climate zones, on their way to the colder and safer marshlands of the northern Hemisphere. While crossing the Hindu Kush mountain range, they often land on the Chitral River and its tributaries, where hunters build artificial ponds on dry riverbeds to ambush the ducks from close range.

Mastuj, a lush green valley located a two hours jeep drive from Shandur on the Chitral side, is a critical location for two major flyways: one along the Chitral River and into Afghanistan and then

The Horned Lark, a Eurasian bird, photographed during its migration over Shandur.

back down via the Kabul River into the Punjab; and the other up over the Shandur Pass to the Shimshal Valley and from there into Kashmir.

For centuries, Shandur Lake served as an important high-altitude wetland, a safe place for rare ducks like the Gargany Teal and the Ruddy Shelduck. Wildlife specialists who visited the lake in the 1970s witnessed ducklings in Shandur in the month of July, although they were not able to identify the species. This evidence of breeding by a migratory population of ducks in Pakistan is rare. Back in the 1970s, however, the polo tournament was not held on a regular basis and moreover, there was no permanent rest house in Shandur, so bird hunters had nowhere to stay. Shandur Lake's very isolation and location on a high plateau made it an ideal stop over for migratory birds.

In the early 1980s, the Chitral Scouts constructed their permanent rest house, and then the polo tournament became an annual fixture, attracting thousands of people from all over Pakistan in the month of July, the prime breeding season for the ducks. The rest house also made it convenient for hunters to stay overnight in Shandur and then go onto the lake for duck shooting in the early hours of the morning.

Shandur Lake could no longer offer a pristine haven for the ducks and eventually the birds stopped breeding in Shandur, although, there is no scientific proof to support this hypothesis. Wildlife specialist Ashiq Ahmed Khan who has visited Shandur over the past four decades, is of the opinion that either the particular migratory population of ducks which used to nest in Shandur has disappeared over the years due to indiscriminate hunting or that the ducks have abandoned this flyway over Shandur during their annual migration. According to Ashiq Ahmed Khan, 'Local hunters report that the Ruddy Shelduck, which was often spotted in Shandur, has not been seen in the Chitral area for the past fifteen years.'

Over the years, with the development of the polo festival, Shandur has lost its importance as a safe wetland for birds. Migratory birds still visit the lake, however. A recent survey of the bird fauna of the Shandur wetland revealed a total of thirty-nine species. These species include the Gadwall, Mallard, Pochard, Shoveller, Pintail, Gargany, Coot, Grebe, and Eurasian Kestrel. No fish were seen in Shandur Lake, although it has good potential for trout. The most serious threat to the lake is clearly from the large crowds of tourists who visit each July, trampling on the lake's surroundings and discharging large amounts of liquid, chemical, and solid waste into the wetland.

Mastuj, a lush green valley located a two hours jeep drive from Shandur on the Chitral side, is a critical location for two major flyways: one along the Chitral River and into Afghanistan and then back down via the Kabul River into the Punjab; and the other up over the Shandur Pass to the Shimshal Valley and from there into Kashmir.

Realizing that fighting over a territory with such colossal mountains would be impossible, in 1886 the Anglo-Russian Boundary Commission carved out the Wakhan Corridor, presently the border between Afghanistan and Pakistan. Nomads at Bozoi Pass are shown descending into Wakhan.

All photographs of Wakhan courtesy Inayat Ali of Shimshal

From Afghanistan: The Wakhan Wilderness*

The nineteenth century was a period of intense political tension between two major powers in the high mountains of the Himalayan chain in the Wakhan Corridor. Tzarist Russia and the British Raj in India were vying for power and control of the Hindu Kush, Karakoram, and the Pamir Mountains. Realizing that fighting over a territory with such colossal mountains would be detrimental to the interest of either kingdom, the two sides formed the Anglo-Russian Boundary Commission in 1886. This Commission carved out the Wakhan Corridor which now forms the border between Afghanistan and Pakistan. The zone was forced on Amir Abdur Rahman Khan, the ruler of Afghanistan, so that at no point would British India and Tsarist Russia touch the territory. Unbeknown to either side, this landscape of immense beauty forms one of the world's superb natural splendours, with glaciers, rivers, montane valleys, and high plateaus surrounded by peaks that rise over 6,000 metres. From a distance, the serene valleys seem lifeless in the quietude of the towering glaciers shimmering in bright sunlight. A closer view of the valleys, however, tells a different story. There is abundant vegetation in the meadows and numerous species of wildlife thrive in the valleys of meandering waterways. Among them are the legendary Marco Polo Sheep, Markhor, and the endangered Snow Leopard.

A stupendous range of mountains, named the Pamir Knot by past travellers, separates the Wakhan Corridor from the Northern Territories of Pakistan. These mountains stretch for about 200 kilometres from east to west, allowing human and wildlife access only through one mountain pass, the Boroghil, reaching an elevation of 3,800 metres. This pass forms the major flyway for birds on their migration from northern breeding grounds to the south in autumn and in the opposite direction in spring. The migration of large flocks of birds, especially the noisy waterfowl, is an exhilarating sight. Large flocks of waterfowl such as Grey Heron, Bar-Headed

*By Khushal Habibi.

Goose, Ruddy Shelduck, Mallard, Widgeon, Teal, Pintail, Shoveller, and Goosander hover low over the horizon in an aerodynamic V-shape with flock members constantly changing positions with the lead birds. Their quacking noise reverberating in the solitude of the tranquil valleys tells of the coming of a change in season. The poignant Pashto poet, Khushal Khan Khattak, has phrased the trepidation of migration in these words:

> *If you want to know separation's shock,*
> *Look at the stray heron that has lost its flock.*

Like most other passes in the Hindu Kush range, the Boroghil is closed during the winter months to human traffic, and when the skies are overcast in fall and spring, the birds are unable to cross the pass also. As a result they settle in the wetlands of the mountain valleys awaiting for the weather to clear and use the Zor-Kol (4,100 metres) and Chaqmaqtin (4,000 metres) lakes in the Big and Small Pamirs as resting sites, and the Ghizar wetlands, south of Boroghil, in spring when they are en route to their northern breeding grounds.

The presence of a large number of waterfowl invokes the hunting instinct of hunters who come out after the quarry. In the past, these hunters used primitive weapons and hunting methods and the number of birds taken was low. With the availability of shotguns, the birds are at a disadvantage as large numbers

These mountains stretch for about 200 kilometres from east to west, allowing human and wildlife access only through one mountain pass: the Boroghil, reaching an elevation of 3,800 metres.

*Chaqmaqtin (4,000 metres) Lake in the Little Pamir is a resting site for migratory birds
as they cross into Pakistan's Shandur wetlands, south of Boroghil.*

Buzkashi, whose players are known as 'chapandaz', in the Pamir Plateau and the northern steppes of Afghanistan, is a roughshod game that has its origins with the Mongol hordes of Genghis Khan, who overran Afghanistan in the thirteenth century.

are hunted indiscriminately. Such practices are ecologically unsustainable and has resulted in the demise of the wildlife of the area, in particular the migratory waterfowl. Species such as the Ruddy Shelduck, a large deep buff-brown duck found along the shores of inland lakes, lagoons, rivers, and streams in these mountains, is decreasing in the region and has been listed as an endangered species. This species, which used to breed in the Shandur region two decades ago, has stopped using the Ghizar wetlands. Hunters in Ghizar, Hunza, and Chitral now complain of a lack of waterfowl in these mountain valleys not realizing a drop in the number of migratory birds, using the corridors, is a result of their own unsustainable hunting practices. Since enforcement has not been practical or effective on either side of the border, both in Pakistan and Afghanistan, the best solution to reverse such negative trends is to start an intensive environmental education program which highlights the benefits of sound and balanced ecological system and its importance to human society.

Good horsemanship in these highlands is of paramount importance. Horse breeding in Badakshan and other northern provinces of Afghanistan is a matter of prestige and pride. Buzkashi players, known as *chapandaz*, are revered much like cricket players. They are the icons of praise and a good *buzkash*, literally meaning a goat dragger, is much praised in the circles of society. The horses are thoroughbred and from infancy are trained to stand up to the rigours of the game.

While in Shandur a regulated tournament of mountain polo takes place with teams in uniforms and horses with knotted tails jolt for a wooden ball from one end of the field to another, *buzkashi* is another story. The *buzkashi* played across the border in the Pamir Plateau and the northern steppes of Afghanistan, is a roughshod game which is said to have its origins in the Mongol hordes of Genghis Khan who overran Afghanistan in the thirteenth century. One of the tactics of the Mongol raids was to grab a sheep or goat by its horn or neck hairs and carry it away—jumping fences and obstacles in their path. From this barbarous beginning, a rough and dangerous game evolved which gained popularity and has

Given the nature of nomadic travelling patterns, this lifestyle is in balance with what the land provides.

Winter in the Wakhan is unforgiving with temperatures falling twenty to thirty degrees centigrade below freezing point.
If the nomads are not prepared for the winter months, survival of the entire family is at threat.

now become the national game of Afghanistan. Played in natural settings, one rider competes with another or a set of riders. The game begins when the carcass of a freshly beheaded goat or calf is placed in a pit or circle at one end of the field. At a signal, the mounted horsemen manoeuvre, jostle, and fight to grab the carcass, break free of the melee and race towards the opposite side of the field, pass the goat around a flag and then try to bring it back to the circle. Rivals try to snatch the goat away and head for the circle. In the ensuing chaos, the goat lands on the ground several times, and the riders once again circle to pick it up. The rules are fluid when played by throngs of *chapandaz*, but there is an unwritten code of conduct among the horsemen to minimize injuries. In essence the game seems like a huge stampede of a thundering mounted army with just as many spectators on horseback joining the chase.

The Wakhis are people of mixed origin who have established permanent settlements along the Wakhan River. The Wakhis, together with other neighbouring Ismaili groups, the Ishkashimi, Zebaki, and Shignani occupy the most remote lands of the Hindu Kush Range in Afghanistan and the Karakoram Mountains in Pakistan. The Wakhis are a small and distinct linguistic and cultural group who are adherents of the Ismaili faith. They practice agriculture and cultivate crops of wheat, barley, and peas. In addition to their agricultural practices, the wealthier among the population own large numbers of sheep and goats, which are annually pastured in the high valleys of the Big Pamir during the spring and summer months. Here they come in contact with the pastoral Kirghiz. Living in a harsh environment, with little agricultural land, these Ismaili communities lead a subsistence existence in the most remote parts of the high mountain valleys. Their life is full of turmoil. Most of the people are engaged in agriculture and can barely grow enough food for the family. The climate of Wakhan is extremely harsh as a result of which only one crop can be grown during the short growing season which starts in May, after the snows have melted. Winter in the Wakhan is unforgiving with temperatures falling twenty to thirty degrees below freezing point. If the Wakhis do not prepare for the winter

A Kirghiz nomadic family in their yurt home in the Pamir Plateau.

*Early in the morning, a Kirghiz nomadic woman milks her yak:
an animal uniquely adapted to high mountain environments.*

months, survival of the entire family can be threatened. They spend the long winter in their villages reminiscing the past and tell stories of fairies and jinns and the legendary Kaaf mountain, access to which is blocked by demons which have to be subdued to gain admittance to this land of utmost beauty and riches.

Like the Boroghil and Shandur Passes, the Lowari Pass, which is over 3,000 metres in elevation, is a formidable barrier to human travel. In winter months the Lowari receives several metres of snow. All traffic to Chitral stops except on days when the weather is clear and it is possible to fly to Chitral. Hence there is always a backlog of people both in Chitral and Peshawar wanting to fly either way. The people of Chitral have found a solution to the menace of the Lowari by crossing over into the Konar province of Afghanistan. They travel north to Arandu and cross over back into Pakistani territory with the Lowari and its snows behind them. This seasonal detour has now turned into a commercial route by means of which supplies now reach Chitral throughout the year. The barrier imposed by Lowari shows that with goodwill and cooperation insurmountable obstacles can be overcome.

The nomadic herdsmen of Afghanistan, known as *koochi*, move as a group from summer to winter pastures. Traditionally the eastern *koochis*, mainly belonging to the Ghilzai tribe move into the lower plains of Pakistan. These people are in constant search for new pastures, and over the centuries they have established set routes. In parody during spring they chase the snow into the mountains, while with the onset of winter the snow chases them to the lowlands. In the early 1960s, political tensions between Pakistan and Afghanistan caused a complete shift in their migration patterns, and the *powindah*, as the nomads are known in Pakistan, had to readjust their migration routes. In the 1980s, the *powindah* once again started using their traditional winter grazing areas.

The life of the nomad has been much romanticized in poetry and literature. A closer look at their lifestyle, however, suggests a different story. Constantly on the move, the nomads are travellers the year round. Their only abode is their woollen tent. Living under the stars they begin their day at two in the morning when they start their seasonal trek. They travel for about eight hours before the day becomes hot by the sun. When on the move, they pitch a temporary tent and then prepare their meal. At times they are in hostile territory and have to guard their possessions. Their fierce dogs serve this task well for no stranger is allowed in the midst of the camp without the consent of their master. When on the move, scouts stay ahead of the caravan to look for pastures to which they can guide the family caravan.

Given the nature of their travelling patterns, this nomadic lifestyle is in balance with what the land provides. Their alternate grazing pattern, in use for centuries, ensured the use of the mountain pastures on a sustainable basis.

The mountain wetlands are not only ecologically important in the arid landscape of the Hindu Kush and the Karakoram Ranges but are vital to the survival of the human communities which have grown around these wetlands. They are prime areas of biodiversity, and besides providing breeding and nesting grounds for waterfowl and serving as resting areas for migratory birds, they are also the main source of fish and crustaceans. They attract large numbers of rodents and insects thus forming 'islands' of biodiversity in a parched mountain landscape. Plant communities thrive around these wetlands, and the abundant sedge meadows are the major attractants of some of the most fabulous species of spectacular ungulates such as the Marco Polo Sheep, Himalayan Ibex, Urial, Blue Sheep, Musk Deer, and the legendary Markhor. These herbivores, in turn, are preyed upon by the leopard, Snow Leopard, caracal, and the ubiquitous wolf. Their presence creates a well-balanced ecological utopia.

Introduction

Most people conceive of Pakistan as an arid, irrigated landscape densely inhabited by people. Yet, in the Northern Mountains of the country lies what is sometimes known as the third pole; a region so full of frozen masses of ice as to be identified as a major site of glaciers in addition to the North and South Poles. The confluence of the Hindu Kush–Karakoram–Himalaya, in the region have made this part of the Himalayan arc particularly concentrated in glaciers, an anomalous geology with some of the largest concentrations of high mountains caused by rapidly growing uplifts of the earth's tectonic plates.

Rather than being static masses of stored water, glaciers are moving and dynamic entities affected most by the energy they receive through temperature changes across both long and short time horizons. During normal annual cycles, glaciers are responsible for a major source of meltwater that flows into the nation's river systems, and allows the production of food and energy for millions of people. The dramatic landscapes of the mountain complexes of Pakistan are shaped by the processes of glacier movements over millennia, as well as some of the catastrophic water releases from time to time. Thus we call them guardians of time: geologic formations that withhold the power to shape change in the physical landscape around us.

We describe a village in Chitral district where a recent glacial lake outburst flood has destroyed half the village, capturing the witness of residents who heard and saw the cataclysm that a moving glacier causes and how the event cannot be foreseen or predicted either through science or traditional knowledge.

In the art of growing glaciers, we recount the traditional methods and lore of how villagers have seeded glaciers to grow so as to provide life-giving water to expanding mountain habitations. The account of a glacier 'mating' still exists in living memory in the mountains, and one of the country's eminent fiction writers is quoted recounting this activity in her short-story.

(facing page)
In north Pakistan lies what is sometimes known as the third pole; a region so full of frozen masses of ice as to be identified as a major site of glaciers in addition to the North and South Poles. We call them guardians of time: geologic formations that withhold the power to shape change in the physical landscape around us.

All photographs in this chapter are courtesy of Ghulam Rasool, unless otherwise stated.

*Glaciers are a major source of meltwater that flows into river systems and
allows the production of food and energy for millions of people.*

Finally, the two-hundred-year-old history of Western knowledge about glaciers is recounted, beginning with colonial annals of distinguished explorers, and today is the subject of scientific debate and controversy. Glaciers are particularly sensitive to global climate change, and whilst scientists agree that the eastern Himalayas are melting faster than previously noted, the western Himalayan arc is behaving differently. The complexities of these inscrutable glaciers are still not clear to scientists, and the methods and debates of geologists and mountain specialists continue even as communities live adjacent to these life-giving water masses that can potentially devastate all life around them.

The western Himalayan arc region is considered one of the best laboratories
to examine relationships between climate, glaciations, and landscape evolution.

The Third Pole

Contrary to the common image of Pakistan as an arid, irrigated country of plains and deserts, it is thought that outside of the North and South Pole, the belt of high-altitude mountains located in Pakistan's territories is among the most glaciated regions in the world. The Hindu Kush–Karakoram–Himalayan arc amounts to a great mass and collection of glaciated ice, and is sometimes referred to as the Third Pole.

This region is also considered one of the best laboratories to examine relationships between climate, glaciations, and landscape evolution. The dynamics are not well-understood by geologists; current knowledge establishes that these are young mountains formed in the Holocene era of geologic time, which is an epoch thought to coincide with the earliest humans. The mountain origins are thought primarily to be controlled by uplifts of the tectonic plates of the earth in this region, and the landscape we see today is an outcome of a combination of surface processes such as erosion caused by glacial movements and uplift mechanisms marked by great oscillations and glacial advances in geologic time.

Of particular interest are the Karakoram Mountains which are sometimes described as 'an anomalous topography' home to more than sixty peaks exceeding 7,000 metres and one of the most rapidly rising parts of the earth with an uplift rate between two to six millimetres annually. It is for this reason that the K2, the second highest mountain peak in the world may well grow higher than Everest in the eastern Himalayas in the near future.

The Karakoram Range is also one of the foci of concerns of climatic change. In contrast to the eastern Himalayas, which have an average ice cover of 8 to 12 per cent, the Karakoram consist of 28 per cent glaciers on average, with patches of glaciation up to 48 per cent such as at Siachen. Thus scientifically, the region is considered highly sensitive to changes in temperature and precipitation patterns.

With so much water stored in glacial masses of ice, changes in water release patterns from these stores have powerful effects on local and regional levels. The Indus Basin Irrigation Systems relies heavily on the run-off generated by the melting of snow and ice in the Upper Indus Basin areas; it is estimated that 40 per cent of the average annual flows of the Indus comes from glacier basins. Hydropower energy provides 28 per cent of the installed power capacity of the country, most importantly from the two large dams at Tarbela and Mangla. Future dependence on dam construction as a source of water and power will increase, and so enhance Pakistan's dependency on annual water inflow into reservoirs to meet rising demand. For this reason, glaciers have assumed a central role in concerns of Pakistan's water management.

The fourth working group of the Intergovernmental Panel on Climate Change points out 'widespread mass losses from glaciers and reductions in snow cover over recent decades are projected to accelerate throughout the twenty-first century, reducing water availability, hydropower potential, and changing seasonality of flows in the regions supplied by meltwater from major mountain ranges including the Himalayas and Hindu Kush where more than one-sixth of the world population currently live.'

According to glaciologist Geoffrey Archer, the Karakoram glaciers and permanent snowfields provide long term storage for precipitation, buffering unreliable annual rainfall and providing constant water input into the Indus system.

Among key drivers of water flows are the high-altitude upper Karakoram catchment areas where large glaciation feeds the Hunza, Shigar, and Shyok rivers. The annual run-off from these is strongly dependant on energy inputs into the glacier systems through seasonal temperature variations. Due to climate change, it is thought that the annual runoff from these high-altitude catchments might change. If the snowline in these areas recedes then a higher direct run-off in winter will happen and a decrease of water level in summer will occur, a time when it is most needed for irrigation in the growing season.

There is fear that drought years will become more common as thermal and hydrological patterns alter; these fears are the source of much debate centred on the unusual behaviour of the glaciers of the Karakorams.

Perhaps the foremost geographer to have studied the region's glaciers, Kenneth Hewitt, emphasizes that contrary to findings of Himalayan glaciology where there is an observed pattern of receding glaciers, the Karakoram glaciers may have started thickening and advancing. He claims that, 'there is no question that today's behaviour is a regionally distinct response to climate change.' He adds that advancing glaciers is also a dangerous situation. Alongside continual research controversy over climate changes in the Karakoram, he highlights that, 'in every valley of the region, farmers tell me the winters have grown shorter in the past couple of decades, there is less snow and more rain.' Agricultural cycles also seem to be changing in concord with these subtle shifts in weather and climate, and as Hewitt points out: 'these are, in fact, more immediate hazards for the mountain communities, than anything that may be happening to the glaciers.'

Glaciers are dynamic landscape features that move slowly, shaping topography over millennia: the 'imprint of glacial action' is part of the landscape itself; glaciers cover nearly one-third of the basin areas of north Pakistan. In their dynamics, glaciers recurrently exhibit short, sharp bursts of events to create lakes, glacial dams, and river floods. Depending on the magnitude of these events, they can cause destruction to human infrastructure, change the course of river flows, and create new landscape elements such as lakes and valleys.

Let us turn to a particular valley in upper Hunza called Shimshal as a case of a habitation that abuts glaciers of significant size. All of the events described above have been experienced in the living memory of the people of Shimshal. Locally, glacial events are perceived to be beyond the agency of human power, but are understood to be connected to global climate change trends. As protection against their dangers, the weather specialist of Shimshal informs that the community 'has been given [through the Ismaili teachings] special religious instructions for practices and prayers' against glacier induced destruction.

The Indus Basin Irrigation Systems relies heavily on the run-off generated by the melting of snow and ice in the Upper Indus Basin areas. For this reason, glaciers are a priority concern of Pakistan's water management.

Glaciers recurrently exhibit short, sharp bursts of events to create lakes, glacial dams, and river floods.

Shimshal is one of the most highly glaciated areas of Pakistan containing six major glaciers that are masses of frozen ice dating back to the late ice age. The Khurdopin and Shujerab glaciers are two glaciers known by the community as posing imminent danger—and they are being constantly watched: Khurdopin as a moving glacier that has recurrently caused overflow of Shimshal Lake and rise in river; and Shujerab as the cause of landslides that causes damage to animals and people.

In addition to local knowledge, scientific data has focused on Khurdopin Glacier since 1884 with the latest recordings till 2000. Although this is a short time span for glacial dynamics, consolidated analysis has been provided by Hewitt. The Khurdopin Glacier is the third largest glacier in the Hunza region, after Hispar and Batura, and is known to have created numerous and destructive floods on record. These floods are known as Glacial Lake Outburst Floods (GLOFs). The succession of glacial surges by Khurdopin Glacier from 1905 to 1908 have caused large scale devastation in Hunza with bridges lost and the creation of a lake some 3.5 by 1.5 kilometres in size.

In 1884, the largest known devastation that occurred due to the Shimshal River was caused at Ganesh-Altit. In 1905, a wave of 10 metres above summer maxima was recorded in the Hunza Valley, a rise in 9 metres in twenty-four hours at the Indus at Bunji, and a peak discharge at Attock of 386,000 cubic feet per second with an increase of 105,000 cubic feet per second in twenty-four hours.

In 1963, a GLOF event in Shimshal is recorded in flood reports by Hewitt as, 'a large dam drained catastrophically beneath the glacier; enormous devastation in Shimshal. The flood wave built up again in the lower Shimshal gorge. Bursting forth as a huge, destructive wave into the Hunza Valley, it wiped out over one-third of Passu's occupied and irrigated agricultural land. Further damage [occurred] as far as Gilgit.'

These unknowable patterns of glacier behaviour are a reminder of how fundamentally and suddenly glaciers move and change human habitations, rivers, and landscapes. Despite almost two centuries of scientific enquiry, they remain inscrutable guardians of locked up energy that can be released in short bursts of time. Thus we call them the guardians of time.

Sonoghure village at 2,000 metres has a glacier above the village. Half the village now lies buried beneath fifteen feet of boulders and stones that came down in a massive flood caused by a melting glacier in June 2007.

Paradise in Peril

'The noise was deafening, louder than thunder,' recalls Sher Afzal, a resident of Sonoghure village located in Chitral District in the Hindu Kush mountain range. 'It was past midnight and I ran out with my four children and saw the water gushing into the village. We barely managed to escape. We saved ourselves but lost everything else.'

Sonoghure, was once considered to be the most beautiful village in Chitral. It was referred to as 'paradise on earth' in local songs and folk tales for its picturesque orchards, towering poplar trees, and old, stone houses built on a cliff overlooking Chitral River just across the road to Mastuj.

Half the village now lies buried beneath fifteen feet of boulders and stones that came down in a massive flood caused by a melting glacier located above on a mountain in June 2007. The incident was barely reported in the national press, let alone the international media. Mountain communities living in these remote valleys, who depend upon their goats, fruit trees, and terraced agricultural fields for their livelihoods, are not responsible for climate change. Yet they are amongst the first victims of rising temperatures in the region.

Luckily the villagers of Sonoghure managed to survive the flood which lasted three days without any loss of lives, although they lost most of their livestock. What happened in Sonoghure is not an isolated incident, and scientists are saying these floods are due to slow melting and that more than twenty-five glacier floods have been recorded in Bhutan, Nepal, and Tibet since the 1930s; more will likely occur as climate change progresses. According to Pradeep Mool, a remote-sensing expert, working for the International Centre for Integrated Mountain Development in Kathmandu: 'the low-altitude and mid-altitude glaciers are shrinking the fastest. As valley glaciers retreat, glacial lakes are forming. This is a new phenomenon that is being observed at elevations between 3,000 metres to 5,000 metres.'

Sonoghure is located at around 2,000 metres and the hanging white glacier above the village is clearly visible on the rugged

A Glacial Lake Outburst Flood occurs when lakes formed by the rapid retreat of glaciers increase in volume, finally bursting out of the unconsolidated moraine and ice which dams them.

mountain behind the village. There is a dark cavern at the bottom from where the water rushed out, cutting a ravine below. The technical term for the flood that devastated the village is a Glacial Lake Outburst Flood (GLOF). These occur when lakes formed by the rapid retreat of glaciers increase in volume, finally bursting out of the unconsolidated moraine and ice which dams them. This leads to a sudden discharge of huge volumes of water and debris. The catastrophic flooding downstream damages forests, orchards, houses, and infrastructure.

The villagers say they did not see a lake form but observed water spurting out of the ground near the glacier shortly before the massive flood hit the village. The lake must have formed underneath the glacier, which is located at around a two-hour hike up the mountain called 'Pal'. The glacier is still quite large and there are fears that it might burst again, since the heavy snowfall in Chitral in the last few winters has fed the glacier.

Around 112 households out of a total of 330 in the village were destroyed by the flood. GLOFs end up covering the area with boulders and sand; unlike landslides which bring down mineral rich silt which is good for planting, the land affected by a GLOF event cannot be used for agriculture again.

'The flood destroyed not only our fields and homes, but also the village hospital, schools, roads, water pipelines, electricity poles and swept away three bridges on the river that connected the village with the main road,' says Sahib Faraz, one of the local villagers who helped with the rescue efforts. 'We lost walnut and mulberry trees that were over 1,000 years old. Each household had an orchard and we could sell one sack of walnuts for around ten thousand rupees to the local tradesmen. Now we are taking loans and struggling to get by as daily wage labourers. We can't even afford to send our children to colleges and universities for further studies.'

Although it has been eight years since the disaster, the villagers say that the government is still not helping them rebuild their infrastructure, including the bridge that connects them to the main road to Mastuj. Given their limited understanding of climate change, they attribute the catastrophe to God's Will. 'Perhaps we made some mistake and did not make God happy,' says seventy-three years old Bulbul Zar, who cannot remember such a flood in his lifetime or his father's. A large swathe of boulders and sand still cuts through the village, a stark reminder of the GLOF that roared down the mountain changing their lives forever. Throughout Chitral District, villagers look up at their valleys' glacier-clad mountains, and fear that what happened in Sonoghure could happen to them as well.

The confluence of the Hindu Kush–Karakoram–Himalaya is concentrated in glaciers, an anomalous geology caused by rapidly growing uplifts of the earth's tectonic plates.

Today, high above the Bindu Gol valley in Chitral, near Mastuj, lies the white, man-made Bindu glacier.
Locals recount that it is an artificially seeded glacier that was made in 1840 by their forefathers.

The Art of Growing Glaciers

For centuries the villagers of the Hindu Kush and Karakoram Mountains have had to live in the shadow of their large glaciers; the rivers of ice that provide them with precious meltwater for their terraced fields and fruit orchards. As their populations expanded and moved into more remote valleys where there were no glaciers close by and the snow-melt was not enough to provide water for crops, villagers began to 'grow glaciers' in these dry, high-altitude regions. According to local legend, 'male' and 'female' glaciers were combined to grow glaciers on the sides of mountains so that villagers could benefit from their offspring. In fact, in some valleys growing glaciers in this manner became crucial to survival.

Today, high above the Bindu Gol valley in Chitral, which is located on the way to Mastuj, lies the white, man-made Bindu glacier. 'This was an artificially seeded glacier that was made in 1840 by our forefathers,' explains Siraj, a local villager. 'There was no water in our area and so they decided to grow a glacier a hundred feet above the settlement.' He says there is no one left in the village with the knowledge of how to grow a glacier today, but in the past they heard that 'cow manure, salt, and straw' was combined to plant the glacier. 'It took three years for it to grow into a glacier. Now it is so vast that there was a huge glacial lake outburst flood in August 2010 that ended up blocking the Chitral River down below the valley for eight hours, since so much debris came down with the flood.' Siraj has climbed up the Bindu Glacier; he says it takes around ten hours to reach the top of the glacier and that it is located in a part of the mountain that receives no sunshine until noon.

The indigenous knowledge of how to seed glaciers began to die out in the Hindu Kush and Karakoram Mountains and today only a few villages still have glacier-growing elders. Glacier growing or grafting used to be a traditional skill of the mountain farmers of this region where it was used for irrigation purposes as far back as or since the nineteenth century. The technique has been described in detail by Lieutenant David Lockhart and Robertson Lorimer, the former a member of the British Indian Army, the latter a noted linguist of the 1920s.

So how does it work? Local tradition believes that there are two types of glaciers: 'male' glaciers are covered in soil or stones and hardly move, while 'female' glaciers are whiter, grow faster, and yield more water. Tradition also dictates that in order to grow a glacier, you need equal amounts of both types of glaciers. Glacier growers sometimes 'graft' glaciers together by selecting an already existing patch of glacier, digging deep to expose 'male' ice, and adding 'female' ice to it, along with large boulders and rocks. Over time, snowmelt trapped in the young glacier freezes, creating more ice. When the newly created frozen mass is heavy enough, it begins to creep downhill, forming a self-sustaining glacier within four years or so.

In a short story published in *Granta* magazine in 2010 called 'Ice Mating', writer Uzma Aslam Khan describes the process in detail:

> *The hole had been shielded by wheat husks and walnut shells. In winter, the covering would be removed so the snow could collect over the two ice blocks—a male, a female. After five winters, the couple would begin to creep downhill, growing into a natural glacier, free of the cultivating hands of men. Freshwater children would spring from her womb providing the village with water to drink and to irrigate their fields... The marital bed— the hole covered in shells and husk—had been dug into the side of a cliff as carefully selected as the bride and groom. Only this side of the mountain attracted the right length of shadow for the snow to hold for ten months, 14,000 feet above sea level. The porters had heaved the ice on their backs the entire way. ... They tossed the male in first. Whooshoo! Whooshoo! A loop of air seemed to dance right back up the hole and circle around again, inside my chest. The female was released on top, falling without a sound.*

Hermann Kreutzmann, a glaciologist at the Freie University of Berlin in Germany, witnessed a 'glacier planting' ceremony in Hunza, near Gilgit, in 1985.

> *It seemed very plausible to me to search for a specific location at the appropriate altitude with a tolerable temperature regime and to place ice there,' he said. As ice can absorb and retain water, he reckoned that 'a substantial amount of ice in a proper location might indeed augment water supplies.*

The practice of breeding glaciers has been on the decline. But rapidly vanishing valley glaciers and water springs have provided a new impetus for drawing on the traditional wisdom of mountain communities. The Pakistani Government is now funding glacier growing efforts and they have reported increased water flows in many of the villages that have undertaken these projects. Generally ice is grown in areas prone to ice accumulation, namely on slopes above 4,500 metres in north-west facing cirques of steep cliffs, and often atop already advancing glacier slopes. Across the border, in the arid region of Ladakh, civil engineer Chewan Norphel, now called the 'Ice Man of Ladakh' has been successfully creating artificial glaciers and bringing water to parched villages.

Today researchers in Europe also wonder whether to augment the existing cultural methods for growing glaciers with techniques from modern research. In some ways, breeding glaciers might become a revitalized fusion of science and tradition; a twenty-first century way to adapt to the challenges of climate change in the high mountain ranges.

Only 20 of the 54,000 glaciers in the Hindu Kush–Karakoram–Himalaya ranges
have been measured. Shown here is the Baroghil Teer-e-Das glacier in Chitral.

The Glacier Controversy

Pakistan has some of the world's longest glaciers such as the massive glaciers of Baltoro and Biafo that together stretch for hundreds of kilometres in the Karakoram mountains and can even be seen from space. Although the world's glaciers have slowly been retreating since 1850, the end of what climate scientists refer to as the Little Ice Age, those in the Karakorams have not melted as quickly as those in other mountain ranges, and in fact, they might be putting on mass. Scientists say this is mostly due to the Karakoram glaciers' location in elevations higher and colder than many other glacier systems.

Although often regarded as part of the Himalayas, the Karakorams are technically a separate chain that includes K2, the world's second-highest peak and spans parts of China, India, and Pakistan. It is home to most of the heavily glaciated area outside the planet's polar regions. Due to its altitude and ruggedness, the Karakorams are much less inhabited than the Himalayas further east. European explorers first visited the Karakorams early in the nineteenth century, followed by British surveyors starting in 1856. The Muztagh Pass was crossed in 1887 by the expedition of Colonel Francis Younghusband, and the valleys above the Hunza River were explored by General Sir George Cockerill in 1892. Explorations in the 1910s and 1920s established most of the geography of the region.

However, there is very little historical information about the glaciers located in these mountains to predict their future. The data that does exist consists mostly of physical measurements taken at the glaciers' most accessible spots — their lowest ends. Glaciers are dynamic and routinely grow in some areas while shedding ice in others. The lower segments are more prone to change due to the higher temperatures associated with lower elevations, making the measurements taken there less reliable. Remote sensing technology has allowed researchers to measure glaciers over larger areas in recent years, but there isn't much historical data to provide a long-term picture.

The Karakoram glaciers have not melted as quickly as other mountain ranges and might be putting on ice mass. Scientists say this is mostly due to their location and elevations, higher and colder than many other glacier systems.

The response of the Hindu Kush–Karakoram–Himalaya glaciers to global warming has of course been a controversial topic ever since the 2007 report of the UN's Intergovernmental Panel on Climate Change (IPCC) was found to have contained the erroneous claim that ice from most of this mountain region could disappear by 2035. The IPCC report stated that the greater Himalayan glaciers 'are receding faster than in any other part of the world and, if the present rate continues, the likelihood of them disappearing by the year 2035 and perhaps sooner is very high if the Earth keeps warming at the current rate.'

There was an outcry when the mistake was detected and the IPCC had to eventually retract the claim. Today there is a broad agreement among scientists that, aside from the Karakoram enigma, the region's glaciers are losing mass. But the melt seems to be occurring more slowly than previously feared and the glaciers of the Hindu Kush–Karakoram–Himalaya Mountain Ranges are expected be around for the next couple of hundred, if not thousand, years.

The region's glaciers provide a vital water source, acting as giant water tanks, for more than a billion people living below in the basins of Asia's mighty rivers such as the Indus. However, insufficient on-site measurements, few high-elevation weather stations, rugged terrain and remote location of glaciers, and sensitive political situations have made both access and

Crossing the Gird base Glacier in 1860s, European explorers first visited the Karakorams early in the nineteenth century, followed by British surveyors starting in 1856. (By permission of The British Library, 752/12(138) Hazara.)

international collaboration difficult to research this region. The region's glaciers have not been studied as well as the glaciers of the Alps or Scandinavia, where every large glacier has been measured and researched.

In 2011, the Kathmandu-based International Centre for Integrated Mountain Development (ICIMOD) published extensive inventories calculating the glaciated and snow-covered area of the Hindu Kush–Karakoram–Himalaya region, based on satellite remote sensing imagery. This was followed by two studies in 2012 calculating glacial mass changes for the mountain chains. One study by US researchers was published in *Nature* in February 2012, and the other by colleagues in France and Norway, was published in the same journal in August 2012. They are regarded as important developments for understanding larger trends in the region.

The studies found that despite the overall loss, no significant mass gains or losses have occurred in the Karakoram region in the early part of the 21st century; the so-called 'Karakoram anomaly' and the most negative rates of mass loss and glacier shrinkage were observed in the south-eastern Himalayas and the Jammu & Kashmir region. According to ICIMOD,

> [T]ogether, these papers provide a better picture of the regional differences in the rates of glacier change in the greater Himalayas. These differences reflect both the range of factors that affect an individual glacier's response to climate change, and the method of detection used. Surface mass balance observations, for example, tend to be more negative than those derived from repeat elevation measurements, while regions influenced by Westerly circulation patterns exhibit large variations in rates of mass loss. Using remote sensing products, an analysis of glacier area and length changes over the past 30 years showed the most extensive glacier reductions in the south-eastern Himalayas.

Kenneth Hewitt, the geo-scientist at Wilfrid Laurier University in Waterloo, Canada, who has been conducting expeditions to K2 for several years to measure large glaciers like Baltoro, has observed five glacier advances and a single retreat in the Karakorams. According to Ken Hewitt, 'Nowhere in the upper Indus Basin do you have the collapse of glaciers like in the Alps. They are actually holding their own or growing. They could well be growing because of climate change. The summer weather is cloudier and there is more snowfall.' Hewitt pointed out that this may be a temporary phenomenon and that there was a serious need to look closely at what is happening. Pakistan's Water and Power Development Authority (WAPDA) is now starting to operate field stations to monitor these glaciers. 'Growing glaciers are not always good news,' explains Hewitt. 'Surging glaciers are dangerous because they store water. The Hunza River has declined by 20 per cent due to the advance of glaciers in the area. These glaciers are storing ice ... This is a different problem and needs to be investigated.'

The effect of global climate on losses or gains of glacier ice is of crucial importance because so many people living downstream depend on the glacier-fed rivers for their water supply. In the Indus River Basin, for example, glacier-melt accounts for about 30 per cent of the river's water supply. Satellite observations are beginning to make the picture clearer, but there is still a continuing need for information. To establish the consequence of glacial melt on the region's local water supply, not just satellite data analysis but assessment on the ground is required. Only twenty of the fifty-four thousand glaciers in the Hindu Kush–Karakoram–Himalaya ranges have been measured on the ground for 'mass balance', or net gains against net losses, with even fewer measured annually. With only a decade of detailed measurements at hand, much more investigation of these massive glaciers is needed to understand how these vast rivers of ice will impact the region's water supply in the years to come.

General Acknowledgements

The Authors jointly wish to thank our two donors. The Rockefeller Foundation's Bellagio Centre Residency Program that allowed Mehjabeen and Rina to complete the manuscript in an enchanted palace and gardens on the shores of Lake Como. The Heinrich Böll Stiftung Pakistan has twice awarded grants to this work over a decade. Initially, Dr Angelika Koster Lossack, Regional Director HBS, sanctioned a grant for completing fieldwork and library research. Recently, Marion Regina Muller, Country Director HBS Pakistan, approved a grant for all the post-publication materials for the book.

The gamut of institutions that facilitated this work include: WWF-Pakistan and the Pakistan Wetlands Project for all our gear, GIS maps, logistical assistance, and unpublished reports; the Houbara Foundation for hosting the desert research; the Frontier Works Organization for hosting our Makran coast field work; the Uthal Coast Guard for our Hingol work; the Jiwani Conservation and Information Centre; the Sindh Wildlife Department for our work on Lungh; the Pakistan Ornithological Society for our work on Rangla; and the Himalayan Wildlife Foundation for our research on Deosai.

Key enabling colleagues and friends include: In Pakistan, Brigadier Mukhtar Ahmad, Vaqar Zakaria, the late Hussain Bux Bhagat, Colonel Shams, Majors Tahir and Shahid, Toby and Nudrat for lending us their library, Tanya Majeed, Amra Ali, Kamil and Khawar Mumtaz, Nadia Ghani at OUP; Roland and Sabrina Michaud in Paris; John Colvin in the UK; and Paul Hawken in California, US, for their reviews and tender encouragement.

Personal Acknowledgements by Lead Author, Mehjabeen Abidi-Habib

I wish to honour my elders who each passed away in the years that this book was under preparation. Within the protective sanctuary of our collective home, I flourished within a harmonious joint-family. My mother-in-law, Miriam, showed me how to make writing flow in the two chapters of this book that she proofed for me while writing her own manuscript on an artist-friend's life. My mother, Jolly, helped me with the first school essay that earned full marks at age 15 years. Herself a star investigative reporter of the nascent 'Jordan Times', she told me to describe places with one's heart on one's sleeve: an experience as delectable as puff pastry. My father, Asghar Ali, whose poetic imagination and adamantine mind set the bar so high that I unlearned perfectionism and got on with the work of writing anyway.

My brothers living far away also need thanks. Yaver listened to our field adventures as he drove me through the streets of London and told me to make them part of this story. Asad slipped me assorted books on how to write; the most unlikely one helped.

In Lahore, my husband Ali and daughters developed a recipe of support with ingredients of wonderment, healthy leg-pulling, and just a pinch of pride. I must thank both my daughters for putting up with the most unconventional example of motherhood in their peer group and still always coming home to roost. Without Ali, the knight in shining armour of Pakistan's nature conservation movement, I wouldn't even have tried this.

Personal Acknowledgements by Technical Author, Richard Garstang

I wish to recognize the constructive role played by the WWF-Pakistan leadership team, especially Syed Baber Ali, Brigadier (retd) Mukhtar Ahmed, Ali Hassan Habib, Anwar Nasim, and Mohammad Younas during the late 1990s and onwards. This empowered Pakistan to successfully raise international support for environmental research and management in a highly competitive global setting. Together these individuals created an enabling environment in which sound, sustainable environmental conservation initiatives could develop. I also wish to extend my personal appreciation to scores of rural Pakistanis who provided scientists and economists with vignettes of the raw data necessary for building effective wetlands management models.

My own perspective in this endeavour was immeasurably enhanced by insights from my wife, Shadmeena Khanum, my sister-in-law Tajmina, and her husband, Hassan Bandaripour.

I am fundamentally indebted to professors Jan Heeg, Gordon Maclean and Jurgens Meester of South Africa's University of Natal, under whose classical tutelage in Biology I received my B.Sc., B.Sc. Hons., and Masters degrees.

Personal Acknowledgements by Assistant Author, Rina Saeed Khan

I would like to thank my late father, Mohammed Saeed Khan, a retired fighter pilot and an avid hunter in his youth who travelled all over Pakistan, for inspiring me to help write this book. He often described those far off places to me, but they always seemed so remote and inaccessible until I began my own voyage of discovery, and what an amazing country Pakistan has proven to be.

In order to complete the research for the book we had to travel from the Makran coast up to the Shandur Pass in jeeps often on our own, and my father was always supportive of me travelling to such isolated and occasionally dangerous places. Although he constantly worried for my safety, he would still encourage my research and travels; he had flown over most of these areas in a plane or trekked there in his youth so we would enjoy sharing stories upon my return to Lahore. The book has been almost a decade in its making, and I'm sorry he passed away last year in December 2014, before he could see it published. He would have been so proud.

I would also like to thank LEAD-Pakistan for providing me with the invaluable training on environment and development, spread over 18 months back in 2002, that added much needed scientific input into my writing for the book. My journalistic writing skills were actually honed earlier during my years as a features editor at Pakistan's first independent English news weekly, *The Friday Times*. I am eternally indebted to my editors at TFT, Najam Sethi and Jugnu Mohsin for giving me the space and support for becoming the writer I am today.

References

Introduction

Hawken, P., *Blessed Unrest: How the Largest Social Movement in History Is Restoring Grace, Justice, and Beauty to the World* (Penguin Books, USA, 2008).

MacGregor, N., *A History of the World in 100 Objects* (Penguin Books, London, 2010).

Section 1: The Coast

Chapter 1
Jiwani: Freshwater's Fringe

'China Set to Run Gwadar Port as Singapore Quits', Syed Fazl-e-Haider for the online *Asia Times*, 5 Sept. 2012.

Coastal Lagoons to Alpine Lakes: A Guidebook to Pakistan's Significant Wetlands (WWF–Pakistan, 2012).

From Memory, Autobiography of Sir Firoz Khan Noon (Ferozsons, 1966).

Gwadar: Integrated Development Vision, published by the Government of Balochistan and IUCN, 2007.

'Inside Balochistan' by Declan Walsh, *Guardian*, 29 Mar. 2011.

Observations on the Waterbirds of Jiwani Wetland Complex, Makran Coast (Balochistan), Attaullah Pandrani, Syed Ali Hasnain, Syed Ali Ghalib and Ejaz Ahmad, WWF–Pakistan, Jiwani, WWF–Pakistan, Karachi (SAH, EA), Consulting Ornithologist, Karachi (SAG).

Chapter 2: Hingol: Temples of Creation; Chapter 3: Sonmiani: A Nurturing Existence

Abidi-Habib, S. M. (2010), 'Ecosystem People: Resilience and Adaptive Capacity in Community Based Conservation in Pakistan', PhD thesis, Government College University, Lahore, Pakistan.

DGIS—WWF Tropical Rainforest Portfolio 1996–2001, WWF Zeist, The Netherlands.

Floyer, E. A. (1882), *Unexplored Balochistan* (London: Butler and Tanner, 1996).

Hughes-Buller, R. (1907), 'Makran and Kharan', *Baluchistan District Gazetteer Series*, Volumes VII and VII A.

Marco Polo, *The Travels* (1289), Translated by Ronald Latham (Great Britain: Penguin Books, 1958), 293.

Personal Interviews (Apr. 2002), WWF Karachi, Dr Ejaz Ahmed, Fayyaz Rasool, Ali Hasnain with members of the Sonmiani community including the Sonmiani and Damb CBOs, and Hayat of Bhera village.

Pottinger, Lt. Henry (1816), *Travels in Belochistan and Sind*, East India Company, A. Strahan, London.

WWF (2001), 'Amazing Mangroves', *Natura* Magazine, Vol. 28, Issue 4, Lahore.

Section 2: The Deserts

Chapter 4
Lungh: Royal Shikar to Sanctuary

Abidi-Habib, S. M. and A. Lawrence (2007), 'Revolt and Remember: How the Shimshal Nature Trust Develops and Sustains Social-Ecological Resilience in Northern Pakistan', *Ecology and Society* 12 (2).

BirdLife International, 'East Asia-Pacific Flyway', http://www.birdlife.org/flyways/asia _pacific/index.html

Encyclopaedia Britannica (1974), Vol. 7, 'Oxbow Lake', 645.

Lungh Lake Wildlife Sanctuary, Sindh Wildlife Department, http://www.sindhwildlife.com.pk/protectedareas /lunghlake.htm

Khan, Najam ul Huda, *Coastal Lagoons to Alpine Lakes: A Guidebook to Pakistan's Significant Wetlands* (Pakistan Wetlands Programme, WWF–Pakistan, 2012).

Miscellaneous Wet-Notes from the Pakistan Wetlands Programme.

Shirazi, S. A. J., 'International visitors: birds come fling in', *Wildlife of Pakistan*, http://www.wildlifeofpakistan.com/PakistanBirdClub/birdcomeflyingin.html

Simmon, Robert, 'Indus River, Pakistan', NASA Images, Earth Observatory, http://earthobservatory.nasa.gov/IOTD/view.php?id=43890

UNEP–WCMC (August 1990), 'Langh (Lungh) Lake Wildlife Sanctuary', http://www.unep-wcmc.org/sites/pa/0937v.htm

World Migratory Bird Day 2010, Pakistan Wetlands, http://pakistanwetlands.org/webpages/migratory%20bird.html

Chapter 5
Rangla: The Incidental Marsh-Moorland

Ahmad, Hilal, Nizar Ahmed Dar, Jan Mohammad, and Umargani Jammal Mohammad (2012), 'Ethnobotany, Pharmacology and Chemistry of Salvadora Persica', *L. A. Review*, Research in Plant Biology, 2 (1), 22–31.

Ali, Zulfiqar, Masood Arshad, Salma Umber, Muhammad Akhter (2003), 'Biological and Socioeconomic Analysis of Central Indus Wetlands Complex, Pakistan', Abstracts of 23rd Pakistan Congress of Zoology, 72–3.

Chaudhry, Muhammad, Masood Arshad, and Ghulam Akbar (2012), 'Some observations on threatened and near threatened Avifauna of Pakistan', Zoological Survey of Pakistan (21), 65–72.

Hokkaido, 'Threatened Birds of Asia, Marbled Teal', *BirdBase*, http://birdbase.hokkaido-ies.go.jp/rdb/rdb_en/marmangu.pdf

Jan, A. U. (1993), 'Wise use of wetlands in Asia', Proceedings of the Asian Wetland Symposium-International Lake Environment Committee Foundation (ILEC), (Kusastu, Japan), 82–4.

Khan, A. A. and A. Dasti (1999), 'Community based management plan for Rangla Wetland Complex', UNDP, Birdlife International, OSP and Department of Punjab, Wildlife and Parks.

Khan, A. A. and M. A. Shah (1993), 'Marbled Teal Breeding in Punjab, Pakistan', *The Bulletin of the Threatened Waterfowl Specialist Group*, 4:7.

Khan, Najam ul Huda, *Coastal Lagoons to Alpine Lakes: A Guidebook to Pakistan's Significant Wetlands* (Pakistan Wetlands Programme, WWF–Pakistan, 2012).

Marbled Teal, *Marmaronetta angustirostris*, Birdlife, http://www.birdlife.org/datazone/speciesfactsheet.php?id=467

Miscellaneous Wet-Notes from the Pakistan Wetlands Programme.

Rafique, Dr Muhammad (2010), 'Fish Baseline Studies of Rangla Wetlands Complex, Muzaffargarh, Punjab, Pakistan', Pakistan Wetlands Programme, pakistanwetlands.org/reports/Rangla%20lake (2), pdf

Sheikh, Kashif, Aleem Khan, and Saji Nadeem (1998), 'Wise use: A traditional approach to utilize the resources of Rangla Wetland Complex'. *Pak. J. Ornith* 2, 62–72.

Ibid.

Chapter 6
Cholistan: Driven by Movement

Akbar, Ghulam, Taj Khan, Naseeb, Mohammad Arshad (August 1996), 'Cholistan Desert, Pakistan', *Rangelands*, 18 (4), 124–8.

Akhter, Rubina and Muhammad Arshad (2006), 'Arid Rangelands in the Cholistan Desert, Pakistan', Article Scientifique, Secheresse, 17 (1–2), 210–7.

Chari, Pushpa (April 2010), 'A Personal Odyssey', *The Hindu Magazine*, http://www.thehindu.com/arts/magazine/article391403.ece

Danino, Michel (2010), *The Lost River: On the Trail of the Sarasvati* (India: Penguin Books, 2010).

Facing Water Challenges in the Cholistan Desert, Pakistan: A WWDR3 Case Study', Waterwiki.net,http://waterwiki.net/index.php/Facing_Water_Challenges_in_the_Cholistan_Desert,_Pakistan:_A_WWDR3_Case_Study

Gobole, Sanjay (January 2011), 'Cholistan, the Inside Story', http://panunkashmir.org/blog/pilgrimage/cholistan-the-inside-story

Kaul, Vivek (2010), 'Sarasvati: Tracing the Death of a River', *Daily News* and *Analysis*, http://www.dnaindia.com/blogs/post_sarasvati-tracing-the-death-of-a-river_1581502

Lawler, Andrew (2007), 'Climate Spurred Later Indus Decline', *Science*, Vol. 316, No 5827, 978–9.

Lawler, Andrew (June 2008), 'Unmasking the Indus'. *Science*, Vol. 320, 1276–81.

Lawler, Andrew (April 2011), 'In Indus times, the river didn't run through it', *Science*, Vol. 332, 23.

'Living Desert, Special Issue' (2001), *Natura*, A Quarterly Magazine of WWF–Pakistan, Vol. 28, Issue 3.

Madella, Marco and Dorian Q. Fuller (2006), 'Palaeoecology and the Harappan Civilization of South Asia: A Reconsideration', *Quaternary Science Reviews* (25), 1283–1301.

Osada, Toshiki (2010), 'Environmental Change and the Indus Civilisation', Research Institute for Humanity and Nature, http://www.chikyu.ac.jp/rihn_e/project/H-03.html

'Other Cholistan Forts', Tourism Development Corporation of the Punjab, http://www.tdcp.gop.pk/tdcp/Destinations/HistoricalPlaces/Forts/CholistanDesertForts/tabid/267/Default.aspx

'Silk Road', Alpine Trekker and Tours Private Limited, http://www.alpine.com.pk/silkroad.php

Wink, André, *Al Hind: The Making of the Islamic World: The Slave Kings and the Islamic Conquest 11th-13th Centuries*, Vol. II (India: Oxford University Press, 1997).

Section 3: The Mountains

Chapter 7
Deosai: Water in the Wilderness

Abidi-Habib, S. M. (2000), Transcript of interview with Vaqar Zakaria and Dr Anis ur Rehman, and Hagler Bailly, Pakistan Offices, Islamabad.

Ali, Salim (1946), 'An ornithological pilgrimage to Lake Mansarowar and Mount Kailas', *Journal of the Bombay Natural History Society*, Vol. 46, 286–308.

Chan, Victor (1994), *Tibet: A Pilgrimage Guide*. Moon Publications, Chico WWF Pakistan (2011), Management Plan for Deosai National Park, WWF–Pakistan Gilgit.

Deosai National Park, Wildlife of Pakistan, http://www.wildlifeofpakistan.com/ProtectedAreasofPakistan/Deosai_NP.htm

Hedin, Sven (1935), *A Conquest of Tibet*, Reprinted in 1994 by Book Faith India, Delhi.

Haughton, Henry L. (1907), 'Diaries of Henry Lawrence Haughton Indian Army, 1907–1912', 5 volumes, including Gilgit-Baltistan, Bristish Library Mss Eur D531, 100 et al.

Khan, Najam ul Huda (2012), *Coastal Lagoons to Alpine Lakes: A Guidebook to Pakistan's Significant Wetlands* (Pakistan Wetlands Programme, WWF–Pakistan, 2012).

Khrinley, Dungkar Losang (2002), *Dungkar Tibetological Great Dictionary* (Beijing: China Tibetology Publishing House), Tibetan Language.

Lihua, Ma (2007), *Ma Lihua's Account of Going Through Tibet: The Journey West to Ngari* (Beijing: China Tibetology Publishing House), Chinese Language.

Mountfort, Guy (1970), Pakistan's Progress, Oryx, *The International Journal of Conservation*, http://www.oryxthejournal.org/

Rizvi, Ahmer Ali, 'The Giant Bears of Deosai', http://www.wildlifeofpakistan.com/Features/thegiantbearsofdeosai.htm

Schaller, G. B., *Wildlife of the Tibetan Steppe* (Chicago: University of Chicago, USA, 1998).

Wright, B. and A. Kumar (1997), 'Fashioned for extinction: an expose of the Shahtoosh trade', Wildlife Protection Society of India, New Delhi.

Chapter 8
Shandur: A Joust with Nature

'For a Living Planet' (2005), Brochure published by WWF–Pakistan.

Jamil, Rehan Rafey (July 2008), 'Shandur: The world's highest polo ground cleans up its act', Newsletter of the Pakistan Wetlands Programme.

Johnson, David (2006), 'Reflected in Water? Assessing the Impact of the Shandur Polo Festival', M.S. Thesis, University of Oxford/Pakistan Wetlands Programme, 1–28.

Khan, Muhammad Ibrahim and Raja Abid Ali (2005), 'Langer-Shandur Wetland Complex', Information Sheet on Ramsar Wetlands (RIS), Ramsar Convention Bureau, Switzerland.

Khan, Najam ul Huda, *Coastal Lagoons to Alpine Lakes: A Guidebook to Pakistan's Significant Wetlands* (Pakistan Wetlands Programme, WWF–Pakistan, 2012).

Miscellaneous Wet-Notes from the Pakistan Wetlands Programme.

Chapter 9
Glaciers: The Guardians of Time

Abidi-Habib, S. M. (2010), 'Ecosystem People: Resilience and Adaptive Capacity in Community Based Conservation in Pakistan', PhD thesis, Government College University, Lahore, Pakistan.

Andreas, Kääb, Berthier, Nuth Etienne, Gardelle Christopher, Julie, and Yves Arnaud (August 2012), 'Contrasting patterns of early 21st century glacier mass change in the Himalayas', *Nature* 448.

Archer, David R. (2003), 'Contrasting hydrological regimes in the upper Indus Basin'. *Journal of Hydrology*, 274 (S), 198–210.

Bajracharya, S. R., P. K. Mool, and B. R. Shrestha (2007), 'Impact of Climate Change on Himalayan Glaciers and Glacial Lakes: Case Studies on GLOF and Associated Hazard in Nepal and Bhutan', ICIMOD in cooperation with UNEP Regional Office Asia.

Faizi, Inayatullah (2007), 'Artificial glacier grafting: Indigenous knowledge of the mountain people of Chitral', *Asia Pacific Mountain Network (APMN)*, Bulletin 8, Nr. 1.

Gardelle, J., E. Berthier, and Y. Arnaud (2012), 'Slight mass gain of Karakoram glacier in the early twenty-first century', *Nature Geoscience*, published online 15 April 2012, http://www.nature.com/ngeo/journal/vaop/ncurrent/full/ngeo1450.html#/access

Hewitt, Kenneth (2009), 'Glaciers of the Karakoram Himalaya', Un progetto fotografico-scientifico dell' Associazione Macromicro.

Hewitt, Kenneth (2009), 'Understanding Glacier Changes', *The Third Pole*, http://www.nature.com/nature/journal/v488/n7412/full/nature11324.html

IPCC (2007), 'Contribution of Working Group II to the Forth Assessment Report of the Intergovernmental Panel on Climate Change, 2007', http://www.ipcc.ch/publications_and_data/ar4/wg2/en/contents.html

IPCC (2009), Key Scientific Developments since the IPCC Fourth Assessment Report. P. C. o. G. C. Change, Arlington Virginia.

Iturrizaga (1997), 'The valley of Shimshal—A geographical portrait of a remote high mountain settlement and its pastures with reference to environmental habitat conditions in the North-West Karakorum (Pakistan)', *GeoJournal* 42 (2–3), 303–28.

Hewitt, Kenneth (2005), 'The Karakoram anomaly? Glacier expansion and the "elevation effect", Karakoram Himalaya', Mountain Research and Development 25 (4), 332–40.

Hewitt, Kenneth (2006), 'Glaciers of the Hunza Basin and Related Features', in H. Kreutzmann, *Karakoram in Transition* (Karachi: Oxford University Press, 2006).

Khan, A. R. (2001), 'Searching evidence of climatic change: an analysis of hydro-meteorological time series in the upper Indus Basin', Working Paper 23, P. C. S. N. 7.

Khan, S. (2005), 'Glacier Grafting', The Aga Khan Rural Support Programme, Skardu, Pakistan.

Khan, Uzma Aslam (2010), 'Ice Mating', *Granta* 112.

Kuhle, K., 'The past Hunza glacier in connection with the pleistocene Karakoram ice stream network during the last Ice Age', in H. Kreutzmann, *Karakoram in Transition* (Karachi: Oxford University Press, 2006).

Mock, John (1998), 'Tourism in Northern Pakistan's Karakoram and Hindu Kush Mountains', 2009.

'New Studies Provide Insights on Glaciers in the Greater Himalayan Region'. ICIMOD website: http://www.icimod.org/?q=8249

Perkins, S. (2012). 'Renegade glaciers gain ice', *Nature*, 15 Apr. 2012, http://www.nature.com/news/renegade-glaciers-gain-ice-1.10448

Tveiten, Ingvar (2007), 'Glacier Growing—A Local Response to Water Scarcity in Baltistan and Gilgit, Pakistan', Thesis submitted for Masters in Development Studies at Norwegian University of Life Sciences.

Zeidler, Julian and M. J. Steinbauer (2008), 'Climate Change in the Northern Areas Pakistan: Impacts on glaciers, ecology and livelihoods'. G. C. a. I. Centre, Gilgit, WWF–Pakistan, Internal Report Version 1.2.

Index